GOLD! GOLD!

Beginners Handbook:
How To Prospect For Gold!

J. F. Petralia

All rights reserved. No part of this book may be reproduced or transmitted in any form by any means without permission in writing from the publisher, except by a reviewer, who may quote brief passages in a review.

Contemporary photos by author
Design by Carol Hoover
Drawings by Susan Neri
Front cover photo by Ben Benet
Typesetting by Type by Design
Editorial assistance by Jill Applegate Petralia

Acknowledgements: Nineteenth Century photographs and sketches courtesy of:
  Levi Strauss & Co.
  The Bancroft Library, U.C. Berkeley
  Wells Fargo Bank History Room
  California Historical Society
  California Division of Mines and Geology
  Southern Pacific Transportation Co.
  Federal Reserve Bank of San Francisco
  Old Mint Museum, San Francisco

Copyright 1980 by Joseph F. Petralia
Revised Copyright 1982 by Joseph F. Petralia
Library of Congress Catalog Card Number: 81-126200
ISBN: 0-9605890-3-1 Softcover
ISBN: 0-9605890-2-3 Library Edition

Sierra Trading Post / P.O. Box 2497
San Francisco, CA 94126

Printed in the United States of America, 1980

Dedicated to all those whose love of the great outdoors and lure of the past will be expanded in this new hobby.

Hardrock miners at Virginia City Comstock Lode wearing their "Levi's." *(Photo courtesy of Levi Strauss & Co.)*

Contents

CHAPTER I
In the Beginning . . .
The King of Money · Gold: Its Nature ·
Gold Fever · The Rush had Begun ·
The Fever Spreads · Mining Life ·
The Motherlode   9

CHAPTER II
The Pot at the End of the Rainbow
Where to Begin Your Search · In Short ·
Why it is Where it is · Its "Heft" is the
Main Clue · Images: A Nugget's Travels ·
Learn to Read the Stream · Placer
Paydirt · Prospecting for Quartz Veins   23

CHAPTER III
Putting Together your Grubstake
Prospecting Gear   51

CHAPTER IV
Methods: Then & Now
How to Pan · Gold Pan Sieve ·
The Sniffer · Dry Placer Operations ·
Operation of a Sluice · Clean-up ·
Rocker Box · Long Tom · Hydraulic
Concentrator · Dredges · Metal Detectors   57

CHAPTER V
Crescent of Gold
Amalgamation · Staking Your Claim   84

CHAPTER VI
Head for the Hills   91

CHAPTER VII
Nickel Knowledge
Survival Pack · Specifications ·
Troy Weight · Glossary · Mining Supply
Resources · Bureau of Land Management
(BLM) Offices · Mining Supplies ·
Equipment Order Form   95

Chapter I

In the Beginning...

Throughout history the story of mankind and the mention of gold have been closely intertwined. It is almost universally considered to be the symbol of everything that is precious and of enduring value, thereby creating its role as a store of value for individuals and entire nations. The form that it has taken has been multi-faceted: from nuggets, ingots, coins, and idols to the current coin mintage of various countries around the world. It has been highly prized for its own attractive nature as well as its ability to withstand the rigors of time. It has been considered dear both because of the effort required to extract it from nature and its scarcity relative to the other metals of the earth.

Mankind has valued gold since the beginning of civilization. The ancient Egyptians were fashioning artifacts from gold as far back as 4000 B.C. and it is the first element mentioned in the Bible (Genesis 2: 10–12).

> A river flowed out of Eden to water the garden, and there it divided and became four rivers. The name of the first is Pishon; it is the one which flows around the whole land of Havilah, where there is gold, and the gold of that land is good.

Although gold was somewhat scarce in early times, there was enough available to be used in

Down the Sierra Nevadas in 1865 *(Courtesy of Wells Fargo Bank)*

9

both daily transactions and as a medium of exchange. It's been the natural way for man to preserve capital and the fruits of his labor as well as a way to protect him from monetary debasement and uncertainty. It has been continually sought since ancient times, plundered from the ancient civilizations of the Inca and the Aztec, and the primary mover in the westward development of this country. Its recent domestic extraction includes several recent gold rushes, beginning in the early 19th century in the southeastern United States, followed by the great California rush of 1849/50 and by later strikes in the Rockies and Alaska. In spite of this pressure, it is estimated that over 80% of all gold still existing worldwide has not yet been recovered.

The King of Money

In more recent times, gold has been used as a hedge against inflation, particularly in countries outside of the United States. The "hoarding" of various bullion pieces (Napolean Francs, Kruggerands, U.S. Double Eagles) is most prevalent during periods of adverse world conditions and economic uncertainty.

(Courtesy Federal Reserve Bank, San Francisco)

Gold – the ultimate money – Why? Because it is the only monetary asset that isn't someone else's liability. It doesn't represent a promise to pay and it isn't dependent upon the survival of a particular power or group of powers. In a word, it is valuable because it is.
How To Wheel and Deal in Gold and Silver. C. M. Allen. 1974.

Gold: Its Nature

Gold is unique among the metals and is considered noble and beautiful by many people. Among its unique physical properties, gold is the most malleable and ductile of all metals, with the ability to be stretched or drawn. It has been estimated that a single ounce of gold can be drawn into a wire over 40 miles long without breaking. It is also an extremely dense metal having a specific gravity of 19.2 (19 times heavier than water). Gold is an excellent conductor of electricity. Its "nobility" means that no substance that appears commonly in nature will destroy it. It is virtually immune to the effects of oxygen and therefore will not corrode, tarnish, or rust. Caches of coins unearthed after centuries from both sea and land have been recovered as brilliant as the day they were lost.

Besides its physical properties, its luster and deep yellow color have, since its initial discovery, lured men and women with an attraction beyond rational comprehension. It is said that if you stare at gold long enough it begins to glow with an iridescence of its own, drawing the observer into its aura. Many have succumbed to what is commonly referred to as "gold fever."

From 1933 until December 31, 1974, it was illegal for Americans to own gold bullion in the United States. Exceptions to this included certain coins which were considered to be of numismatic

Sutter's Mill

quality and gold in its natural state. The ending of the 41-year ban started among Americans a boom in ownership of gold in its various forms. The anticipation of its legalization toward the end of 1974 caused the price of gold to rise steadily, in part buoyed by foreign speculators waiting for Americans to cash in on their new-found freedom. Subsequent leveling of the market price in the ensuing months left some disappointed with its performance. The artificial rise dissipated and gave way to the increasing pressures of the market place. The economic policies of world governments have brought the price of gold up to its current levels. It has been anticipated that its rightful place in today's dollars is over the $1000 mark, which should be reached by the mid-point in this decade. World events, however, could push the price above this level in a matter of weeks. It is obvious that the opening of this new decade will give birth to a new gold rush which brings us to the subject of this book, namely, how to participate and recover your share of "free" gold as it exists in nature.

Gold Fever!

The so-called rush of '49 began with the discovery in 1848 by James W. Marshall on land recently acquired by John C. Fremont. Captain Sutter, a Swiss immigrant, founded the first inland settlement in northern California and built a fort in what is now Sacramento. Later Sutter contracted Marshall to build a sawmill to assure himself of a steady supply of lumber to continue the expansion and settlement of the interior of California.

In January of that year a dam and race (channel) were built on the American River.

For four months these men washed at Coloma, seeing no visitors, and rarely communi-

Sutter's Mill – site of the 1848 discovery of gold in California.
(Courtesy, The Bancroft Library,)

cating with the fort. The Mill had been nearly completed, the dam was made, the race had been dug, the gates had been put in place, the water had been turned into the race to carry away some of the loose dirt and gravel, and then had been turned off again. On the afternoon of Monday the 24th of January, Marshall was walking in the tail-race, when on its rotten granite bedrock he saw yellow particles and picked up several of them. The largest were about the size of grains of wheat. They were smooth, bright, and in color much like brass. He thought they were gold, and went to the mill, where he told the men that he had found a gold mine. At the time little importance was attached to this statement. It was regarded as a proper subject for ridicule.
Century Magazine, The Century Co., New York. Nov. 1890 – April 1891.

Since none of these men had any extensive background in mining, they were neither very excited nor even sure that the "yellow stuff" he had picked up was gold.

Marshall hammered his new metal, and found it malleable; he put it in the kitchen fire, and observed that it did not readily melt or become colored; he compared its color with gold coin; and the more he examined it, the more he was convinced that it was gold.
Century Magazine, The Century Co., New York. Nov. 1890–April 1891.

Captain Sutter was informed of the discovery later that day and performed his own tests. Convinced, but anxious to complete several building projects he had started, he asked that the news be kept secret for several weeks, but to no avail. Sutter, prior to being fully satisfied about the nature of the discovery, had made certain treaties with local Indians, buying their title to the region around the mill in the event their find proved good. Only the remoteness of the sawmill slowed the swift spread of news about the discovery.

The following month Sutter went to the settlement of Yerba Buena (San Francisco), taking samples of the element with him. While in San Francisco he met Issac Humphrey, an old Georgian miner who examined the pieces and confirmed that he had discovered gold in what would later prove to be very rich diggings. Humphrey outfitted a small expedition which reached the site of Sutter's mill in March, 1848. There he made a "rocker" and commenced work in earnest. After several days prospecting the area, he confirmed that the area was, in fact, rich. The first reports were published in two San Francisco newspapers soon thereafter.

Over the next few weeks and months word of their success continued to reach and excite the civilized world. Young and old were seized by the prospect of picking up a fortune for the seeking of it. Representatives of thousands of families back east and overseas were "grub-staked" to make their fortunes for their families. Coloma, the site of Sutter's mill, became California's first mining camp.

The Rush had begun!

The migration to the mountains began with those already in the province, from the settlements in San Francisco and Sonoma, Monterey, San Jose, Santa Cruz, and as word spread, from the communities in the south. As word reached across the Pacific, vessels loaded with fortune-seekers began arriving from the Hawaiian Islands, from posts in the northwest, and eventually strong interest developed on the east coast.

Some immigrants took the southern route around the Horn, other adventurers forming partnerships and companies, travelled overland to the rich mountain streams of the Sierra. The journey was bewilderingly harder than most could imagine,

Early argonauts join in the "rush."
(Courtesy of The California Historical Society, San Francisco)

and many had to wait weeks for a craft to round the Horn before proceeding to San Francisco. Hundreds could only afford the voyage in steerage class, and some contracted the fevers of the tropical regions never to reach the Pacific shore. The overland immigrants had to suffer the rigors of the elements and the Indians; those departing too late never made it over the winter mountain passes.

The scene upon arrival in San Francisco was one of various languages, talents, and backgrounds of merchants, commoners, and soldiers, most under-equipped and under-supplied but of one common purpose. Most had arrived too late to get into the mountains to do any good that year. The ground swell of news coming from the earlier arrivals out of the mountains served only to excite further those who were waiting. The true rush was to begin in the Spring of 1849.

Living in San Francisco during that winter was expensive because of the numbers of men and the scarcity of supplies. However, where work was available, the pay was high, and payment prompt. This did little to temper the excitement of the time. The impression was that the gold fields would be exhausted in a year or two, and it would behoove those who expected to gain much to be among the earliest in the fields. There was little background for most of the men in this type of experiment and the various patent mining contrivances they had brought with them had to be replaced with the pick, shovel, and pan before they were to depart.

In 1848 the gold hunters didn't need a scientific education. The method of washing gold was then so simple — with the savings of a week's work he could buy the pick, shovel, pan and rocker which were the only necessary tools. The auriferous deposits of the Sierra Nevada was done on the bars of rivers, where the gravel was shallow and rich.
Miners Own Book, San Francisco, Published by Hutchings and Rosenfield, 146 Montgomery Street.

The stereotype of the gold rush was born in the fall of '49 and the spring of '50. Times were free and easy. Most had left both their families and the restraints of contemporary society behind them. It was a generation of bachelors, their families several mountain ranges away. Gambling houses abounded with displays of gold dust, nuggets, and "slugs." The streets were filled with men arriving daily with their pouches full. They parted with the gold freely, as men can when it is easily obtained.

... the testimony of the miniature rocks; the solid nuggets brought down from above every few days, whose size and value rumor multiplied according to the number of her tongues. The talk, day and night, unceasingly and exclusively of "gold, easy to get and hard to hold," inflamed all new comers with the desire to hurry on and share the chances.

History of Marin County; Historical Sketch of the State of California. Alley, Bowen & Co., 1880.

The Fever Spreads!

As word of new discoveries became known, more and more people caught the "fever."

The discovery of these vast deposits of gold has entirely changed the character of upper California. Its people, before engaged in cultivating their small patches of ground, and guarding their herds of cattle and horses, have all gone to the mines, or are on their way there. Labourers of every trade have left their work-benches, and tradesmen their shops. Sailors desert their ships as fast as they arrive on the coast . . . many desertions, too, have taken place from the garrison within the influence of these mines.

At present the people are running over the county and picking it out of the earth here and there, just as a thousand hogs, let loose in the forest, would root up grounds-nuts. Some get eight or ten ounces a day, and the least active one or two.

Guide to the Gold Region of Upper California by William Thurston esq. 1849.

It's possible that a 20th century version of this experience will be relived as the price of gold climbs over the $500.00/ounce mark.

Mining Life

The method of the day was to work in groups or companies of men extracting the easiest and most accessible gold then moving on to richer grounds once the fertile area was depleted. The general theory was that the richest gold was at the source. Therefore it followed that moving further up the gorges and unexplored streams and creeks

Early woodcut.
(Courtesy of Wells Fargo Bank History Room)

would offer the richest finds. Hence, the development of the various camps and towns along the tributaries of the major gold-producing rivers (Yuba, Tuolomne, Feather, American, etc.).

> A turn of the road presented a scene of mining life, as perfect in its details as it was novel in its features. Immediately beneath us the swift river glided tranquilly, though foaming still from the great battle which a few yards higher up it had fought with a mass of black obstructing rocks. On the banks a village of canvas that the winter rains had bleached to perfection, and round it the miners were at work at every point. Many were waist deep in the water, toiling in banks to construct a race and dam to turn the river's

course; others were entrenched in holes, like grave diggers, working down to the "bedrock." Some were on the bank of the stream washing out "prospects" from tin pans and wooden "bateas"; and others working in company with the long tom, by means of water sluices artfully conveyed from the river. Many were coyoteing in subterranean holes, from which time to time their heads popped out, like those of squirrels to take a look at the world; and a few with drills, dissatisfied with nature's work, were preparing to remove large rocks with gunpowder. All was life, merriment, vigor and determination, as their part of the earth was being turned inside out to see what it was made of.

    Marryat, Frank, Mountain-Molehills, London. 1855. p. 234.

The Motherlode

The gold regions of California are generally broken up into three categories: the so-called northern, the central, and the southern district mines (see map). For the most part, these districts parallel the main north-south road through the "mother lode" area named after the region bearing the most extensive concentration of gold.

Highway 49 runs north and south through California in the central Sierras from approximately Mariposa at the southern boundary to Sierra City in the north, a distance of over 175 miles. Colorful towns, whose names stir images of the past, appeared, such as Angel's Camp, Fiddletown, Columbia, Auburn, Rough and Ready, and countless others whose lifetime did not exceed several weeks. Towns sprang up like mushrooms after a spring rain, some with equal longevity. Most of the 'camps' consisted of little more than a series of tents or cabins that were only slightly more durable.

Travelling through the various towns today offers a rich experience and insight into the history

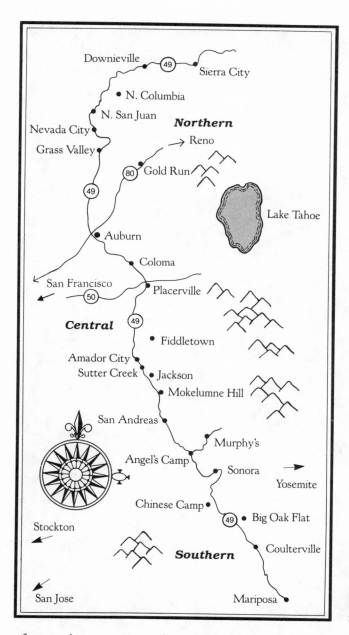

California's "Mother Lode": Route 49 named after the "forty-niners."

of several generations of our past. Many of the towns along Highway 49 are well-preserved. However, many of the lesser-populated and uncommercialized settlements are on side roads easily reached by the family car.

Chapter II

The Pot at the End of the Rainbow

The expression "Gold is where you find it" has been used by generations of prospectors. The reasoning is simple: although there would seem to be technical and physical reasons for recovering gold from the place it should logically rest, that's not always the case.

> Of all metals, gold is, with the exception of iron, the most widely distributed over the earth; but it differs from the latter metal in being present usually in a nearly pure state, but in exceedingly small quantities, whereas iron is abundant as well as generally diffused and is never found unmixed with other substances. Owing to the very minute proportion in which gold is often associated with rocks and mineral substances, it does not generally pay the cost of working; and the districts therefore known as "auriferous" or "gold-producing," in the commercial sense of the term, are not so numerous as the fore-going remarks might seem to suggest.
> *The Gold Seeker's Manual* by David T. Anstead, Professor of Biology, Kings College, London. 1849.

As noted, gold is found in most areas of the world in various forms. The leading commercial producer is South Africa, followed by the Soviet Union, Canada, and the United States. It is found in the mountains, the desert, and in the sea, where it is estimated that six parts of gold are found for

Twentieth Century "Miners": Author (left) working claim on the North Fork of the Yuba River.

23

every trillion parts of salt water. Obviously, in this case, its recovery value is lost unless an economical means of obtaining it is perfected. For the purposes of this handbook, we'll limit ourselves to those areas and methods which you will more than likely use.

The most abundant areas for gold in the United States are located in the Sierra Nevada and Rocky Mountain ranges. However, there have been sufficient concentrations of gold in the southeastern United States (mined during 1830–1840), in the Appalachian Mountains, and in the mountainous regions of Vermont and New Hampshire to warrant increasing interest at the current price levels. Some of the more outstanding states for prospecting include California, Colorado, South Dakota, Alaska, Nevada, Utah, Montana, Idaho, Washington, Arkansas, New Mexico, Wyoming, North and South Carolina, and Georgia.

To better understand where to find gold we should first re-examine its nature. The weight of gold concerns us most. One of nature's heaviest metals, with an atomic weight of 196.967, gold is extremely dense. A cubic foot of gold weighs more than one-half ton, and at $500 an ounce, it would be worth over $9 million. With this in mind, let's look at some of the most likely areas for prospecting.

Where to Begin Your Search

For all intents and purposes, those areas that have produced the best in the past will be the most likely to continue producing. Therefore, we should begin our research with the various public information bureaus for records of recovery in earlier years. The local Bureau of Land Management (BLM) regional office has a number of publications outlining specific areas within your state and

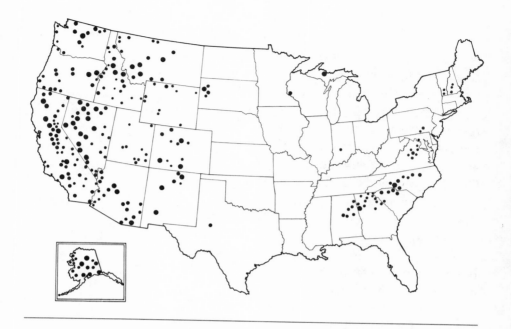

Known gold concentrations in the United States.

county. This department can be very helpful in supplying both topographical maps (at a nominal fee) as well as offering reference material on production levels within their jurisdiction. Your local library can also supply a wealth of information on the gold districts within your region. A check at the State Department of Miners and Geology can provide a geological map of the area showing the mineral deposits.

It should be emphasized that there is a tremendous wealth of information available for the asking. In every instance these departments can provide invaluable aid and cheerful assistance in directing you to the information you seek.

Let's go back to basics using California as an example. . . .

In Short . . .

In ancient times, in what is generally accepted as the Jurassic period, the surface of the central part of the state was uplifted and intruded by molten magma from deep within the earth. In the cooling and contracting that followed many cracks and fissures were formed. Gold and other elements were eventually forced into these openings by the intense pressure of rising hot water and residual gaseous solutions. These gold-bearing *veins,* still deep beneath the earth's surface, were exposed by the erosive conditions of the late Cretaceous period that followed. The streams of the subsequent Eocene epoch ran over and through the metamorphic rock and mineralized zones to form and concentrate the placer deposits. The subtropical climate and other conditions of the period were very favorable to continued disintegration of the "host" matrix.

During the ensuing Tertiary Sierran uplift, the rivers ran basically in a north-south direction. However, the tremendous forces within the earth which caused the uplift and tilting of the region interrupted their flow. Through the passage of time new drainage was formed which cut through the older channels and released the gold deposits. These relatively new watercourses running perpendicular (east-west) to the older Tertiary

The Mexican 'Rastre. Sketch from Hutchings' California Magazine, 1857.

channels are the gold-rich rivers and streams that started all the excitement over a century ago.

> The Tertiary stream gravels, which had long been buried deeply beneath lavas, were exposed by the Pleistocene canyon-cutting rivers. From the dissected portions of the old channels, gold was removed and washed into the newer streams, which concentrated it on their bedrock riffles. The remaining portions of the Tertiary deposits were left with their stubs exposed high up on the intervening ridges. In places, erosion merely stripped the covering of volcanic tuffs, sands and gravels from the bedrock, leaving the channel with its rich gold deposits laid practically bare for the lucky early miner to win.
> Olaf P. Jenkins, Chief, Division of Mines and Geology, Retired. *Geology of Placer Deposits. Special Publication 34. p. 27*

Where the gold is still held in the host rock, it is known as "lode" gold and its extraction is called "lode" or "hard rock mining." The gold occurs in thin veins formed when it, and often granite or quartz, was in a molten state and subsequently forced up from beneath the earth's crust. Commercial operations first have to tunnel into the mountain, or dig a tunnel or shaft, to extract the "ore," perhaps blasting out the surrounding material. The ore-bearing rock would then be crushed to free the gold, using a mechanical device known as a "stampmill." It operated as its name implies: the stamp acted as a giant pestle, rising and falling by means of a cam driven by a power shaft, and crushing the material being fed to it.

In early operations in the past century the action of an "arrastre" was substituted. This was a large round stone fastened to a horizontal wooden arm. The arm was attached to a large vertical center timber which was embedded in the ground and used as a pivot. The movement of the stone, which crushed the ore, was powered by a mule walking in a circle.

On occasion, particularly when the lower grade ore was eventually mined, the arrastre was used in conjunction with the stampmill to further reduce the material and maximize the return of gold. This type of mining is labor-intensive and its overhead closed down many operations. Recently, some of these mines have been reopened because of the current market price of the precious metal.

Why it is Where it is

During the course of erosion, the weathering action will deteriorate the host material of the veins in a gold-bearing region. The action of rain and runoffs will gradually move the eroded gold,

The deposition of gold as it erodes from source to stream placer.

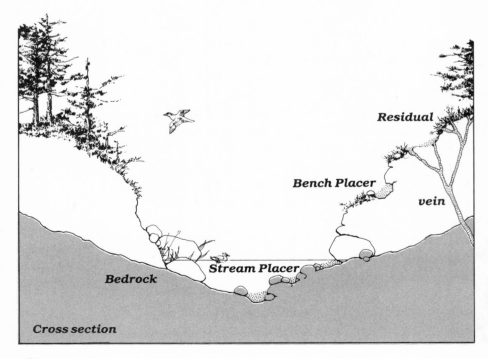

sometimes still clinging to the host-quartz, down the slopes, gullies and drainages of the mountainside. Cumulative deposits will in time come to rest in the nearby creeks and streams. Depending on their location they are classified as either Bench or Stream Placers (illustration).

> In order that a major deposition of gold may occur, there must be an abundance of source material which contains more or less gold and which may be more or less easily eroded. A decayed formation of low-grade material could easily furnish more gold than a hard, higher grade deposit. The decomposed material also supplies more gravel for balanced conditions of stream transportation, providing that overloading or choking is minimized by uplifts or increasing water volume. Plainly, a stream running along a vein system will have a greater opportunity to accumulate gold than one merely crossing it. Bedrock-controlled streams, therefore, provide a maximum contact with source material.

Nov. 1934 paper written by Jenkins, O. P. and Wright, W. Q. California gold-bearing Tertiary Channels, Engineering and Mining Journal, Vol. 135, p. 501.

Types of Deposits

A placer deposit is the formation caused by the natural erosion of lode ore from its original resting place. The heat of summer, the expansion and contraction of surrounding rock and soil, glacier drifts, earthquake faults, winter snows, summer thaws, and subsequent rains, runoffs, and floods all exert pressure to move and break down the gold vein from its point of origin to the resting place where you can recover it.

As noted above, *Bench* and *Stream Placers,* are similar in nature: the former is an ancient deposit of a streambed that has been isolated, i.e., left "high and dry," as the stream eroded its way

deeper into the face of the mountain, or changed direction, stranding the gold above the current water level. These "benches" can be 20 to 100 feet above the river and can usually be spotted by the round "tumbled" river rocks they contain. By visually locating the point at which the bench lies on bedrock and prospecting the material, you'll be working on choice ground.

A Stream Placer, on the other hand, is located within the range of the present level of water. It contains deposits of gold-bearing material either in "gravel bars" (illustration) or other areas, including subsurface, which, for reasons we'll discuss, accumulated the heaviest water-borne elements.

> For the most part, the original source of gold is not far from the place where it was first deposited after being carried by running water.
> Olaf P. Jenkins, "Geology of Placer Deposits," Mineral Information Service, Vol. 17, Nos. 1–9, 1964.

The richest places, however, are not necessarily those closest to the source. This is not a contradiction. The formation of the placer deposits is heavily dependent upon how the placer material is rewashed by natural forces to concentrate the amount of recoverable gold. I'll elaborate further a little later on.

The *Placer* gold we're after is gold in its natural or raw state. It can be found in sizes ranging from finest "flour" to substantial "nuggets" washed or dredged from either active or dry streambeds. Since placer gold is alloyed in nature with other native elements, its purity will usually vary from 700/1000 to 900/1000 fine.

Let's assume that you have researched the various materials available from the Bureau of Land Management, Bureau of Mines, local history from your library, etc., and are now surveying a

River in late summer. Note formation of "gravel bar" on left bank.

particular area. The easiest method for taking in the greatest land area is the topographical map. If you were to zero in on a specific area in the mountains of Georgia, or a river in Colorado, or a stream in Canada, you would be able to purchase from either the BLM or a large map store, the topographical map covering your particular locality. By doing so, you would have a bird's-eye view of the area that would enable you to study the meandering direction of streams, land contour, and stream gradient. What you should be looking for are areas in streams where gold, during its movement, would become trapped.

Let's review a few places you are likely to encounter deposits in your prospecting. A *Residual* or "primary" deposit is one which occurs at the surface of the ground or at the origin or outcropping of a gold vein which has eroded. In order for gold to be released from its original source in bedrock, the encasing matrix (host material) must be broken down through weathering, glacial action, faulting, uplift, chemical disintegration, etc.

> On its way down the hillside gold is sometimes concentrated in sufficient amounts to warrant mining. Such deposits are classified as *Eluvial* placers—they are transitional between residual and stream, or *Alluvial* deposits.
> *Olaf P. Jenkins, "Geology of Placer Deposits," Mineral Information Service, Vol. 17, Nos. 1–9, 1964.*

The secondary types of deposits are "transported" placers and are classified by their distance from their source and location. Therefore a *Eluvial* deposit is one which has moved a short distance from its original location.

Imagine, if you will, the movement from a residual deposit as it travels down the face of a weathering mountain to a stream. Picture a vein of pure gold an inch or so in diameter and several feet long just under the surface of the soil. If you were prospecting electronically, with the aid of a metal detector, and chanced upon the discovery of this vein, in its original location, you would have discovered lode gold. If this gold had eroded, the immediate area would be a residual deposit.

Eventually the placer gold we are looking for will work its way down in the gullies, creeks, streams, and rivers that drain these mountain regions. The type and texture of the underlying rock structures of these waterways will determine the retaining ability of each area.

> A further and very important factor is the ability of the bedrock to hold the deposited gold in spite of the scouring action of the stream at higher water stages. A smooth, hard bedrock is a very poor one for placer accumulations. Bedrock formations which are decomposed or possess cracks and crevices are good, and those of a clayey or a schistose nature are excellent in their ability to retain particles of gold.
>
> *Nov. 1934 paper by Jenkins, O. P., and Wright, W. Q. California gold-bearing tertiary Channels, Engineering and Mining Journal, Vol. 135, p. 501.*

River/stream bottoms which have been displaced to form a "ledge" perpendicular to the direction of the stream can provide an excellent subsurface pocket for accumulation of gold.

Its "Heft" is the Main Clue

Gold, being heavier than any other materials you are likely to find, will be moving very reluctantly in the watercourse of the stream. In doing so, it will generally move in a straight line following the path of least resistance and taking the shortest distance between two points in the stream course.

Bearing these images in mind, let us proceed again to look for a likely place to make our discovery.

One way to approach the study of the topographical map is to look for areas where the stream or river will be making a sharp turn. Gold, moving from its original point of deposit into the stream, will generally move during periods of heavy flooding, high water, spring thaws, or during mountain storms. During this period of heavy water flow there will be several factors that will effect the velocity or speed of the water: namely the stream gradient (the amount of drop or angle of the stream over a particular distance), and the width to contain it. Unless the gold has been freshly deposited, its weight will cause it to quickly settle and embed itself in the various cracks, fissures, roots, and other obstacles in its path as it moves downstream. It will, therefore, take substantial velocity for it to move again (Newton's "object at rest"). Periods of unusually heavy flooding, which occur every few years, tend to rearrange these deposits.

Images: A Nugget's Travels

Imagine standing at a high vantage point during an unusually heavy spring runoff. If you were to drop a relatively large lump of gold, say the size of an almond, in this fast-moving water, how would it travel? If the stream were fast and narrow, with little obstruction, our test lump of gold would probably travel uninterrupted for the full course of the stream. But nature doesn't form streams that are smooth, straight, and narrow. Each turn, new tributary, outcropping, variation in grade, etc., will determine the velocity at each given point.

Again, using the simplified illustration (figure 1), if we were to drop our nugget in a fast-

Figure 1: Boulder or bedrock outcropping at inside bend will cause the formation of a gravel bar—prospect its upstream end.

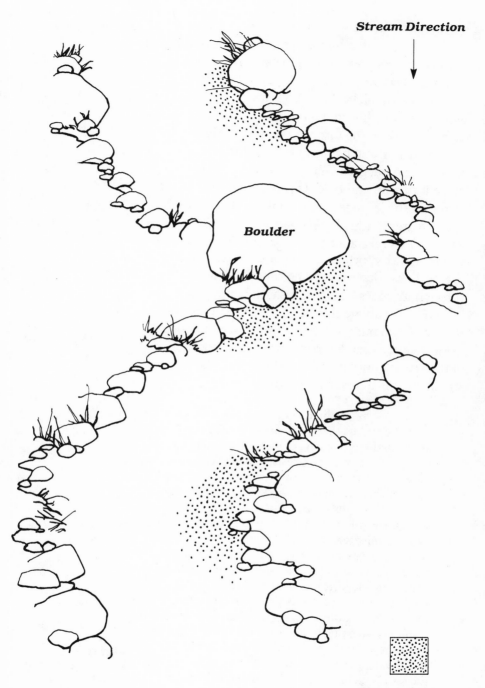

moving, man-made stream with only one obstruction, such as a large boulder in the middle of that stream, the boulder would have an area of still water or relatively calmer water on its downstream side. In all likelihood our lump of gold would be travelling at a more-or-less constant rate carried along by the force of the water. However, upon reaching and being swept around the boulder, the nugget's weight would most likely cause it to stop and hold its position in the calmer water on the lee side of the boulder, there to rest as the water gradually receded.

Let us take another instance using a straight and narrow hypothetical stream of a fixed grade (figure 2) swelled by a spring runoff. Midway through its course imagine a sharp bend. Several hundred feet above this bend we drop in our test nugget. From our vantage point we see the nugget being swept along down the straight, smooth part of the stream until it reaches the point where the stream begins to make its sharp curve.

In a straight channel the current is swifter near the middle than near the sides and is swifter above mid-depth than below. On arriving at a bend the whole stream resists change of course, but the resistance is more effective for the swifter parts of the stream than for the slower. The upper central part is deflected least and projects itself against the outer bank. In so doing it displaces the slow-flowing water previously near the bank, and that water descends obliquely. The descending water displaces in turn the slow-flowing lower water, which is crowded toward the inner bank, while the water previously near that bank moves toward the middle as an upper layer. One general result is a twisting movement, the upper parts of the current tending toward the outer bank and the lower toward the inner. Another result is that the swiftest current is no longer medial, but is near the outer or concave bank. Connected with these two is a gradation of velocities across the

Figure 2: Inside bends have heavier gravel deposits (bars) and often trap gold.

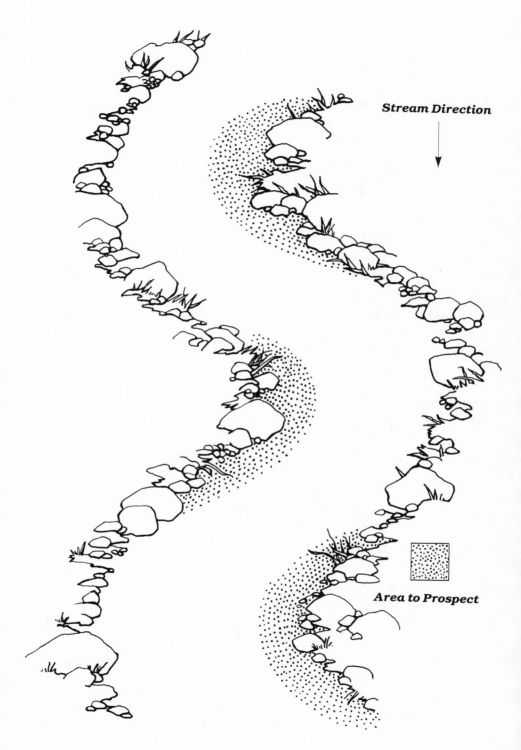

bottom, the greater velocities being near the outer bank. The bed velocities near the outer bank are not only much greater than those near the inner bank but they are greater than any bed velocities in a relatively straight part of the stream. They have therefore greater capacity for traction, and by increasing the tractional lead they erode until an equilibrium is attained. On the other hand, the currents which, crossing the bed obliquely, approach the inner bend are slackening currents, and they deposit what they can no longer carry.

Quoted from G. N. Gilbert, Geology of Placer Deposits, Division of Mines and Geology, SP 34, p. 19.

The water on the *inside* of that curve will be running *slower* than at any other point in the bend, therefore enabling our nugget to resist and settle in the inside curve. Its weight will cause the gold to travel in a straight path following a line of least resistance from inside bend to inside bend.

Another example relating to resistance and deposition of gold would be our stream with equal proportions throughout, but which suddenly increases its width at the midway point (figure 3). Again, our test nugget would travel along quite rapidly, moved by the constant flow of water until it reached the point where the stream widened. The wider point offers the water a chance to slacken, affording the gold an opportunity to settle in the slower water.

The transporting power of a stream is dependent on its velocity, which is a variant determined by the gradient, volume and load. When a stream is overloaded with sediment, the excess is dropped. When it is underloaded, it erodes. When equilibrium has been established, neither erosion nor deposition take place. Gradient, volume and load usually vary in the same stream so that deposition may be going on in one part of its valley and erosion in another.

Figure 3: Placer deposits accumulate in areas where stream velocity is reduced such as in widened areas in streambed, downstream of rapids.

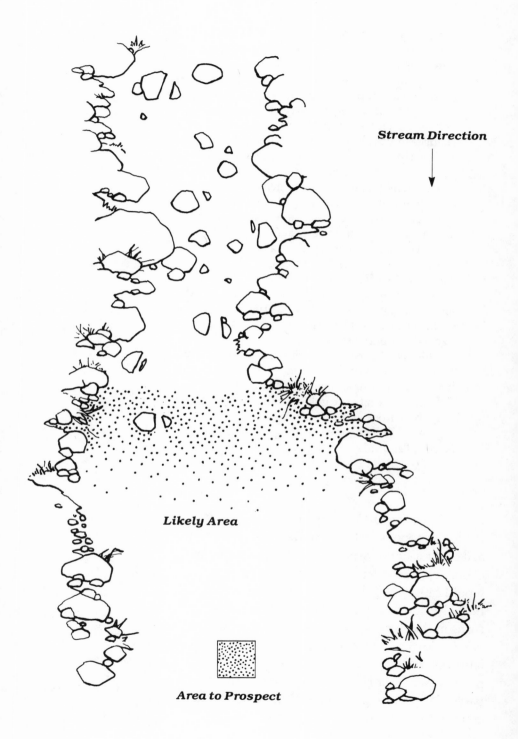

When a stream is eroding, the material within reach of its activity is constantly moved in a downstream direction. All movements of this kind are accomplished by more or less sorting and make for the concentration of the heavier particles.

Deposition takes place in a stream when the velocity is decreased, either by the periodic changes in volume or by a change of gradient. Where there is a change of grade, resulting in diminished velocity, the gold is laid down with the other sediments. It must be remembered, however, that placer gold may find lodgment in inequalities of the bedrock surface where no considerable deposition of detrital matter has taken place, though extensive placers are, as a rule, not formed because of irregularities in the bed-rock surface below. The concentration of gold in river bars is analogous to its deposition in stream beds, for it is dropped where the velocity of the current is checked by the formation of eddies, due to the inequalities of the river floor.

Gold has a specific gravity of approximately six times that of gravel but under water this ratio becomes about nine times. This large gravity difference permits the gold quickly to work its way to bedrock and into crevices.

A. H. Brooks, article on the gold placers of the Seward Peninsula, U.S. Geological Survey Bulletin 328, pp. 125–127.

What we are looking for then, when we study our topological map, aerial photograph, or make a first-hand inspection from a vantage point of our selected stream or river, is anything that would cause the gold to drop or be deposited. Namely, slower water, the inside bends or sharp turns in the river, wide spots in the river which have a slower flow of water, and obstacles such as large boulders, outcroppings of bedrock, deep pools, fissures, or large cracks in the bedrock, particularly those that run perpendicular to the course of the stream.

Deposits are formed by slackened current downstream from obstacles.

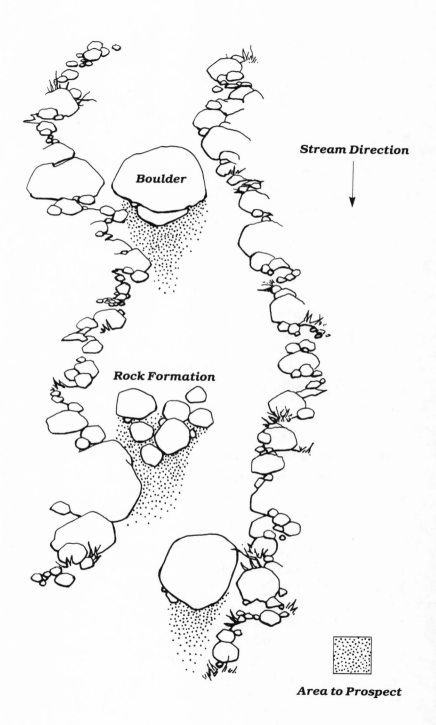

Gold deposited in such locations will remain there until a greater force, i.e., a heavier flood stage with related greater water volume, moves it to a new location. Its resistance will be reinforced by the nature of gold, that is, its density will cause it to settle as deep as it can in the stream bed until it reaches a point of resistance it can not overcome such as bedrock.

The variations on the above set of circumstances are infinite. In the natural state it will be much less common to find only one set of circumstances or obstacles to impede the movement of gold. More than likely there will be several combinations of compound curves, boulders of various sizes, fast and slow water, and several types of sub-surface obstacles to trap our elusive prize.

Rather than list additional simplified examples, your knowledge of the basics will allow you to imagine for yourself the various circumstances that would not allow the gold to escape its transportation through the swift waters of mountain streams.

Learn to *Read* the Stream

It's best if you first study your particular stream selection from both an overview of the area and an on-the-spot inspection at the water level. Walk the stream bed for several hundred yards in either direction. Study the various locations of rocks and boulders. These will range in size from baseballs to basketballs to boulders as big as a house. All have come to rest during past runoffs; like our gold, unwilling passengers riding the force of the flood water. You'll notice that on the inside curves, as we discussed, there will likely be deposits known as "gravel bars." These are excellent sites for the deposit of gold. Look for the perpendicular crevices, particularly if they coincide with the inside curve of the stream or in combination with other obstructions such as bedrock outcroppings. Also, as you make your on-site inspection by walking up and down the stream bank observe not only the present flow of the water, its rapids, and its wide areas, etc., but also consider recent flood waters 5 to 10 feet above normal level, and the less frequent floods – those that occur every 20 to 30 years and which raise the water to a raging river 20 and 25 feet above where you are now standing. There might easily be a large nugget hung up somewhere undiscovered by those looking only at how the river is flowing today.

As mentioned, the various examples I've cited involved the hypothetical situation of introducing our test nugget into the stream to see how it reacts to the flow of water. To better understand where we might find concentrations of gold, which can range in size from almost microscopic grains (flour gold) and matchhead-sized pieces up to and hopefully including nuggets, we will have to bear

Daguerreotype, circa. 1852. Miners working a Long Tom in Auburn Ravine. *(Courtesy of Wells Fargo Bank History Room)*

in mind that the gold we are after did not necessarily erode from the same location or at the same time. The deposits you will be looking for will be those which have accumulated over a long period of time, although, as streams are replenished annually, a considerable amount of gold will have arrived more recently.

What we are looking for are those areas which contain an accumulation of gold, not necessarily overlooked by the original '49ers, but perhaps a deposit accumulated from the last 100 years and perhaps several major floods. This is not to say that the original '49ers cleaned out every area they worked, for their methods of recovery were sometimes crude and haphazard. It's estimated that over one third of the gold they mined during the hydraulic operations was lost during the process as well as a considerable amount lost through the crudely constructed "long toms" that ran for miles along the mother lode terrain.

Another method brought into play in the early mining fields used water in a more aggressive role. *Hydraulicking* was the practice of directing high pressure hoses along the canyon walls to wash down "bench" deposits and other potentially rich gravels into crudely constructed wooden sluice boxes or long toms some exceeding one quarter mile in length. These operations caused severe silt build up in the streams and eventually contributed to damaging build ups in the Sacramento River and in the San Francisco Bay.

You can see the results of using these huge hoses to carve the paydirt today in scarred hillsides that resemble a minor version of the hillsides of the Grand Canyon. It has also been estimated by local miners to whom I've talked that 10 to 15 percent of the placer gold found today is gold that was present and undiscovered during the gold rush of 1849/50. The reason, as stated earlier, is that

Hydraulic Washing.
The scene above represents a company of miners washing down the hill by the Hydraulic process. The water from above being confined in a strong hose, is played through a pipe upon the bank of sand and gravel, with great force and effect. By this process, great quantities of earth are washed down, and passing through a long sluice, the gold is there saved. Sometimes where the gold is very fine, the Guyaskutus is of great value to the miner, saving nearly enough to pay his weekly water bill.

Miners working a sluice box.
(Courtesy of The California Historical Society, San Francisco)

when these miners moved into a rich area they would move on to easier pickings when the original area they were working seemed depleted or was located in an area inaccessible to them (stream bottoms that the early argonauts were unable to divert [Flume] thereby laying it open to mine). I've seen several respectable-sized nuggets that were recently taken in the Yuba River from areas very heavily worked by several generations of miners. This is not something you should expect to find on your first, or even second, try at gold mining, but perhaps the next time, or the next time. . . .

Another point we will discuss in more detail when we talk about the techniques of panning — but one which you should be aware of when making your observations — is the presence, in the likely areas of gold deposits, of old square nails, bits of iron, buckshot, spent bullets, and concentrations of black sand. These elements are proportionally heavier for their size than the rock around them and generally are associated with concentrations of gold. Because of their weight, each settles in areas of lesser resistance, as will our elusive metal.

I strongly suggest spending at least several hours, and ideally several trips, if feasible, during the winter or spring months to observe various areas and subsections during the spring runoff. This is the time to study the flow of the water and to also get a feel for which areas should prove richer than others. When the more temperate weather arrives, you can spend that time recovering gold with a better understanding of where it will most likely be found. On these various shorter excursions you might want to bring along several hand tools to sample each area, thereby eliminating the obviously unprofitable zones. Take notes and pick land marks that you can refer to since an area can look substantially different several months later.

Cross-section of stream bedrock showing cracks, crevices, and depressions that form "pockets" to trap gold.

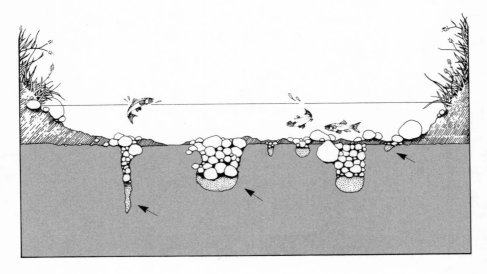

Placer Paydirt

Points to Remember: Look first to the obvious: large cracks, crevices, fissures (particularly those at points where the river or stream will slacken, or wide points, changes in direction, etc.), and the inside bends in the curves of the stream (more than likely associated with deposits of gravel). Observe areas where the river slows down for *any* reason (where the course of the river widens, after a set of rapids, deep pools in the riverbed or where the gradient dissipates thereby slowing flow). Check downstream from intersections of fault lines. Notice sections of river wherever the water current might slacken because of obstructions (boulders, outcroppings of bedrock, etc.). Be on the lookout for concentrations of black sand, old nails, horseshoes, etc. Sample depressions in rocks, particularly those that have many small boulders jammed into the depressions. Check and take a bucket of the material that is caught up in the tangle of exposed tree roots at the water's edge. Make notes of areas in the midsection of the stream where quantities of large boulders have dropped at the point at which the river or stream has widened. Pay attention to any area or condition that would have caused the carrying power of the stream to be reduced. Also look for large boulders that, although "beached" now, were at some time surrounded or completely submerged by a long-forgotten spring flood. Winter or early spring survey trips will reveal mid-stream boulders high and dry and easily accessible by mid-summer. It also pays to check out areas where there's evidence of placer tailings from the "old" days, where miners in the desire to cover as much ground as possible often exhibited "haste made waste" in their recovery methods.

Think about the stream geology we spoke of earlier. Learn to 'read' the stream and the factors that determine the speed and volume of its water. Be on the lookout for areas with black sand along the sandy river banks and areas where the swift water would slacken to allow the heavier gold to settle-out (usually at about a 45 degree angle downstream). Look for areas at the stream bottom where such heavy objects as lost tools, pieces of broken machinery, and other dense objects have come to rest.

In areas containing schistose bedrock, break open the outer slabs with a pry-bar and work out the material contained in the crevices. Check areas where you suspect the original river channel may differ from its present course. Its presence can be detected by the rounded, "river-worn" rocks it contains. Dig beneath the "overburden" wherever possible and attempt to reach bedrock.

Prospecting for Quartz Veins

The following section is reprinted from Special Publications 41, courtesy of the California Division of Mines and Geology.

A gold-bearing vein may or may not be visible on the surface. During its slow process of breaking down, gold becomes scattered in the soil, usually close to or on the bed-rock below the vein. The movement of gold eroded from a vein is like the flow of water.

For example, visualize a small vein of gold-ore occurring on a hillside and running in a direction nearly parallel to the base of the hill. If, at a point about twenty feet below, and on a line parallel to this vein, a number of samples five or ten feet apart are taken by digging down to or nearly to bedrock, they will likely yield gold colors in panning (a color is one visible flake or speck of gold). On a line forty feet below, and parallel to the vein, samples taken in the same manner may also give gold colors, but they will probably be fewer in number to the pan. At sixty, or perhaps as far as two hundred feet below, colors might still be obtained. In searching for gold deposit conditions are reversed: the source is unknown, but the finding of colors is an indication of the existence of a gold-bearing vein at some higher point.

In prospecting a hill, holes are usually dug near its base at intervals of fifty feet or more, and the alluvium near bedrock is panned carefully. When colors are found, the prospector ascends about twenty feet where he digs more holes in a line parallel to the first row. He pans samples on this line and then climbs about twenty feet higher and starts his new line of holes over the point where he obtained his best sample. He is attempting to follow the gold flow to its source by picking up in his pan little specks of the scattered metal. This method of prospecting is called "post-holing" on account of its resemblance to digging such holes to obtain samples.

Prospecting a hillside: Beginning at dot No. 1 a sample was taken, and then again every 50 to 100 feet. At point 8 the best prospect was found. Samples were then taken as represented by + until the crest of the hill was reached; in this case the vein was rich, but did not crop out.

Gold can usually be found on the bedrock of creeks or gullies in gold-bearing regions. To search for gold in a dry creek, find a place in the watercourse where the bedrock is exposed or nearly exposed. Gold lodges under large rocks and in cracks in the solid formation. Find a fracture in the bedrock. Pry it open with a pick or bar. Your pan filled with water should be handy. Lift out the rocks as they are broken, and wash them in the pan, scraping off any adhering clay or sand. Scrape up all the sand from the crevice and place it in the pan. A small paint brush, a spoon and an old table or putty knife are useful in scraping up all the fine sand that might be lodged in a crevice or under a boulder. Scrape the bedrock vigorously and brush up the sand and dust carefully, for the gold flakes sink deeply. Sometimes three or more scrapings from different parts of the creek may be obtained for one panning test. Pan very carefully. If gold is found, ascend the watercourse and continue to pan at spacings of fifty feet or more. When a point is reached where panning does not yield colors, or the amount of colors greatly diminishes, go back to where it was last obtained and "post-hole" on the hillside.

With a pencil and paper one could outline the probable course of quartz or float from a vein and plan the finding of the deposit by tracing the cast-off flakes of gold or pieces of quartz.

Disintegrated quartz that has separated from its vein and become scattered, follows a course like that of the foregoing described gold, only it is likely to travel farther in its downward course. Quartz float can be traced to its source by careful observation.

In testing quartz for gold, pulverize a small amount to the fineness of sand. A small mortar and pestle is necessary for such testing. A large handful of fines will be necessary. Pan very carefully.

Learn to identify quartz. Use a magnetized knifeblade to remove fragments of iron. Have any heavy unknown mineral found in panning identified, but try to learn to identify the nearly black grains of iron oxides. Have quartz that contains fine-grained pyrite or lead minerals assayed for gold.

TOMING.—The above represents three men working with a Tom; two are vigorously picking down and shoveling the dirt into the upper part of the Tom,— and the other is moving it about with a hoe or shovel, to wash it and throw out the larger rocks or riddlings. The gold, dirt and water passes thro' a seive or tom-iron at the lower end into a riffle box underneath, where the gold is saved.

CHAPTER III

Putting Together Your "Grubstake"...

Since you will probably be travelling to the mountains prior to the summer months, at least for your survey trips, there is a good chance you will encounter some adverse weather conditions. It is therefore advisable to carry in your vehicle equipment suited to the locality, such as tire chains, poncho or foul weather gear, insect repellent, and, of course, a good set of boots.

It is usually helpful and a practical safety consideration not to travel alone. This is particularly true in areas where you may have to hike several hundred feet or several miles off the main road in order to reach your chosen location. It is also advisable, if this is the case, to let someone know approximately where you are going and your expected time of return in the event you don't arrive within a reasonable period of time. If you are hiking in, along with your prospecting tools it is generally a good idea to carry a small pouch or kit containing some basic survival items such as a compass, salt tablets, first aid kit, and signal mirror, along with a canteen and perhaps something to eat. All these items can be put into a small backpack or "fanny pack" like those worn around the waist by the ski patrol.

Another point worth mentioning: if you plan to hike down a canyon to either reconnoiter, pan, or even fish, allow yourself enough time to climb

Be aware of prevailing weather conditions!
(Courtesy of The California Historical Society, San Francisco)

out in the daylight. More so in winter, but also in summer months, the light fades very quickly when it approaches sunset. At the end of a long day fatigue can slow you down and make you careless.

This past summer, while camping in the early evening at the top of our claim, we were summoned by a very exhausted and shook-up fisherman. It seems he and a partner had fished too long, misjudged the time, and lost the trail they took down the ravine. The younger man had managed to 'muscle' his way up the side but his partner was stuck on a very sharp, loose wall and caught quite literally between a rock and a hard place. Fortunately, I had a 100-foot section of rope which we were able to anchor to a large fir, lower to the man and pull him up from the tight spot, creel of rainbow trout and all. Starting a half-hour earlier would have prevented the potential tragedy since they had missed the entrance to the trail by less than 100 feet!

Prospecting Gear

A good set of shoes will make the experience safer and more comfortable. If you are operating along a stream bed in the summertime an old pair of sneakers make negotiating the stream safer and easier on the feet. In addition, I've listed below several "sampling" basics:

Gold Pan	Pry Bar
Shovel, Hand-pick, Crevice Tool	Whistbroom
	Old Spoon
Bucket	Several vials to store gold
Coffee can or similar-sized container	Tweezers
	Magnet
Garden Trowel	Small magnifying glass

Prospecting is, by definition, the testing for values in each area of your search for the most likely spot to find gold. The gold pan, a pick, and small shovel are the tools most used for this stage of recreational mining. The gold pan is also a

Early Western scene. *(Photo courtesy of the Donald Simon Collection)*

valuable "sampling" tool that can ultimately lead you to the source of the colors.

The gold pan can either be steel or the plastic variety. Each has its advantages. The traditionalists prefer the metal pan. They claim it is more durable and can be used for other purposes such as a fry pan or wash bowl. The standard size is 16 inches; however, the 12-inch and 14-inch models are easier to handle in the beginning and are fine for sampling an area.

I prefer the new high-impact, plastic variety since it is lighter in weight, certainly as durable under normal use, and can be bought in either black or green to contrast the gold against a dark background. Additionally, the many plastic pans available have "traps" or cheater riffles moulded into them to aid the novice in the recovery of fine gold. Pan selection is a matter of personal preference, but if you're hiking back for any distance you might prefer the advantage of the lesser

Gear–Gear: the basic tools of the prospector.

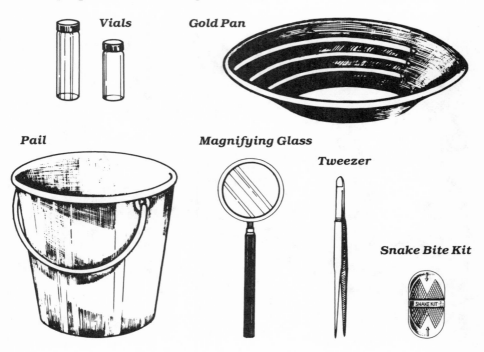

weight of the plastic pan so that you can carry additional equipment of your choice.

If you should choose the metal pan, it is necessary before its initial use to burn or "blue" it. The purpose is to both burn off any oils used in its manufacture as well as darken the surface appearance so that the gold can be seen more easily. The easiest way to burn off the pan is to put it on a stove for several minutes or right into an open fire.

Each type of pan comes in various sizes from 6 to 18 inches; 8 to 12 inches is generally the most popular. An additional advantage of the plastic pan is that if you plan to do any electronic prospecting with the aid of a metal detector, any material added to a plastic pan can be easily checked by passing the detector head over it. Additionally, concentrations of black sands, when dry, can be separated from the gold dust by moving a magnet in a circular motion under the pan.

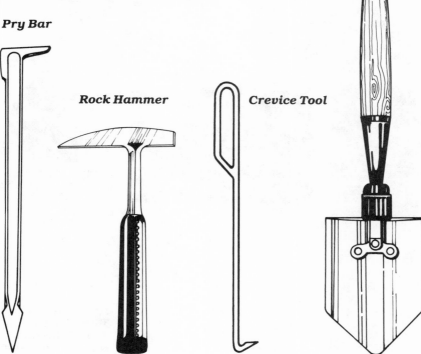

Pry Bar

Rock Hammer

Crevice Tool

Shovel

Hutchings' Mining Methods *(Courtesy of the California Historical Society)*

CHAPTER IV

Methods: Then & Now

We'll begin this section with an overview of the different methods commonly used to recover placer gold. Each of these and their variations were used in the major "rushes" and are depicted in the various illustrations that accompany the text. Where larger companies or groups of men were available, additional techniques were employed to divert the natural flow of the streams or to make use of its power in operating some of the equipment. If water was scarce, canals or flumes were built – in some areas up to several miles long – to bring water to the paydirt.

We'll confine ourselves for the most part to that equipment and manner which you're likely to use or see in operation by weekend miners. Tunneling techniques, stamp-mills, and the like can be pursued at another time.

There are several methods for the recovery of gold, each depending on how deeply committed you want to get and the amount of time and money you wish to invest. The simplest, least expensive, and primary method is *panning*. The object of panning is to separate and concentrate the heavier material by washing away the lighter. The most common panning method is called "wet" panning, in which case a ready supply of water must be available in sufficient quantity. This is generally done at the stream bank.

Panning can also be used as a means to an

end, that is, for sampling a likely area where a more extensive operation will later take place. A pan full of potentially auriferous material can be checked from several different areas along the contour of the stream. More than likely, a spot that shows 'color' at the shallower depths will be richer as you dig deeper to where the gold has settled. This, however, will not always be true. Rich ground will be laid down by the erosion process in layers depending on the conditions that prevailed and in 'stringers' along with the black 'sands'. The idea is to survey and sample likely areas with the pan before settling in for the day or bringing in your heavier tools (pick axe, crow/pry bar, sluice box, etc.).

Since pannng ultimately brings us to the essence of what we're after, I feel it should be described in detail. Unfortunately, as a picture is to a thousand words, so is a demonstration of immeasurable value. Study the text, watch an

Six miners with rocker. Note one miner showing nugget in pan. *(Daguerreotype courtesy of The Bancroft Library.)*

Early engraving

"old-timer" if you can, and practice. You'll catch on in no time. Speed isn't the motive, efficiency is.

How to Pan

Select, through the various observations we have discussed, the most likely areas for an accumulation of gold. Keep in mind that the water action and settlement of the gravel will cause the gold to sink as deep as it can, stopping only upon reaching either bedrock or dense clay. Therefore, when at all possible, dig down to the bedrock and clean out each crack *thoroughly* with the aid of a screw driver or crevice tool. Pry these cracks open and sweep them clean with a small brush or whisk broom. Depressions and holes in the bedrock, which contain stones and rocks jammed into place and held by sand and gravel, should also be pried loose and the material panned. Shake or wash all roots and compacted twigs thoroughly into the pan. Next, with your pan no more than three-quarters full:

1. Settle yourself into a slow-moving section of the stream bed where you can squat and work the

pan in the water. Waders or waterproof boots will insulate your feet from the cold stream. Place the pan completely under the surface of the water by several inches. With one hand, thoroughly mix and knead the material with the water to completely saturate everything in the pan. Be sure to break up all clogs of dirt, especially any clay-like soils. Wash and break up all root material.

2. Next, with the pan still under the water's surface, begin to vigorously shake the pan back and forth and from side to side keeping the contents within the pan. This action will settle the gold and cause the lighter material and muddied water to rise to the surface and be carried off in the stream current. Make sure the water in which you're working is not moving too fast since it would be difficult to regulate the amount of unwanted material washing out of the pan. If you are using a plastic pan with traps or riffles, these should be positioned at 12 o'clock.

PANNING OUT.
The above represents the primitive method of mining. A pan filled with earth is set into the water, and by shaking it from side to side, the dirt is washed out, and the gold gradually sinking to the bottom of the pan, is there saved. This method is still used by every company to wash out the product of the days' labor; while the Chilian or Mexican uses the pan or bowl exclusively.

The "Tennessee Partner" *(Courtesy Levi Strauss & Co.)*

3. The *thoroughly washed* larger rocks can now be removed. Continue shaking the pan to settle the gold.

4. With the pan still submerged and the edge furthest away from you tipped slightly downward, begin a circular or swirling motion either clockwise or counter-clockwise. The material in the pan should be in a state of liquid suspension. This will "float" the lighter material to the top where it will be carried out of the pan. You may rake some of the top-most material off at this point since the gold will have settled close to the pan's bottom. Once again, place the pan under the water and "resettle" the gold by shaking the pan.

5. Continue washing and floating the lighter (blond) sands and other material out of the top of the pan. This action duplicates the action of the stream, 'settling' the gold into the bottom of the pan. Stubborn pebbles should be picked out by hand.

6. Continue the circular motion. Each slight 'thrust' eliminates more material. Proceed

slowly until your proficiency has increased. Stop periodically to resettle the gold in the crease of the pan by rapping the pan's rim two or three times with the palm of your hand. The "traps" or riffles formed in the side of the pan help hold the fine gold.

7. After several minutes you will have only the heavier material remaining in the bottom of the pan. This will usually be the "black sand," a combination of magnetic and nonmagnetic irons and pyrites. Continue to slowly wash this 'concentrate,' being very careful that the fine gold doesn't work its way up the rim. Resettle as necessary. An alternative method to panning down this concentrate is to hold the pan with either one or both hands with its lower edge at a 20–30 degree downward angle, and slowly dip a third or so of the black sand in and out of the water to further reduce the concentrate. Watch for any signs of gold working its way up the rim.

8. When you've worked the black sand concentrate down to the point that only a tablespoon or two remains, remove the pan from the water. Put *just enough* water in the pan to permit the concentrate to swirl. Now, with the pan out of the water and level, start the water moving in a circular motion, spreading the concentrate into a thin layer across the surface of the bottom of the pan and exposing the 'colors!' These flakes can be removed with the aid of a tweezers or with the tip of your finger moistened with saliva. Place the gold into one of the various glass vials that are sold in most hardware and hobby shops. Some miners prefer to fill these small vials with water to help the gold "drop" into the bottle. Excessive black sand that may end up in the vial can be removed by passing a small

magnet along the outside of the vial working the sand up to its opening.

An alternate method entails drying out the concentrate in the sun, then passing a magnet under your *plastic* pan to separate the sand from the gold. If you find buckshot, remnants of square nails, or slivers of lead, the odds are in your favor that you have found a good spot since these heavier items are often associated with concentrations of fine gold. Keep removing and sampling deeper material. Gold is extremely heavy and will settle deeper than other items of similar size.

Don't be in a hurry, especially when you're first learning. It takes at least ten minutes to satisfactorily settle-out a large pan of gravel. If you are doing your clean-up of *concentrate* from a sluice, rocker, or dredge, you'll need to spend even more time. This is the heaviest, richest material, therefore don't risk losing several hours work by saving five minutes with the pan.

Panning is basically a simple procedure, but one with many individual styles. It's best to practice before your initial trip by placing a few BBs or fine buckshot into some test material and pan this in a large wash tub at home. If you can

Early print showing various contemporary mining methods. *(Courtesy of the Dale Chappell Collection)*

find all of the 'simulated' nuggets in the remaining concentrate, you will have achieved enough proficiency to feel confident that you won't lose the heavier gold.

While knowing how to pan is basic and necessary for almost all other phases of prospecting, it is not generally used to recover significant amounts of the metal. There are easier and more efficient methods of doing so.

Gold Pan Sieve

The *Sieve* is a very useful device for increasing productivity while panning. Most of those I've seen and used are made of light-weight, high impact plastic and, depending on the size of the gold pan, will fit inside or over it.

In use, gravel is shoveled into the sieve and pan combination. Both are held together, placed in the stream, and shaken back and forth for a minute or so to "sift" the gravel. This allows the finer material to pass through the mesh (usually three-eighths inches) and into the pan. The sieve is then checked for nuggets and the debris is discarded. This action classifies the gravel, eliminates the need to pick out the stones one at a time, and cuts panning in half. The material remaining in the pan is then washed in the usual manner.

In *all of* the techniques employed to mine gold, *the amount of the metal recovered is directly proportional to the volume of material processed.* In the case of panning, which is relatively slow, the sieve enables the user to process considerably more gravel in less time and with less effort. The sieve or a coarse screen can also be adapted to sort material prior to its introduction in the hopper end of a sluice or rocker-box.

Modern gold pans are made of high impact black plastic that show gold at a glance and contain built-in riffles and gold trap. *(Courtesy of 49er Products)*

The Sniffer

A device sometimes called a gold sniffer or suction gun can be helpful when attempting to clean out a pocket mixed with water and gravel. The nose of the device is placed into the loosened material in the crevice and the plunger withdrawn, creating a suction which fills the canister with water, sand, and gold (see illustration).

Take the material you plan to process from as close to the bedrock as you can dig. The bedrock poses an impermeable layer through which the gold cannot settle. Pockets, cracks and crevices should be *thoroughly* cleaned out. These are natural gold traps. The more compacted the stones and gravel, the greater the likelihood of holding the eroded metal.

The canister contents are then panned.

Gold suction gun or "sniffer." Sucks up water, sand and gold.

Winnowing.
(Courtesy, California Division of Mines and Geology)

Dry Placer Operations

Drywashing, or dry panning as it is sometimes called, is done as the name implies in an area where no water is available at the time. It can be either desert or an area having seasonal runoff. The locations for the latter are generally in dry placer areas of desert stream beds. These streams are rarely dry year-round and the formations to look for here are the same as for active streams. Here again, gravity is the main force for separating the gold from the other materials. There has been a considerable amount of gold recovered in the desert areas but it rarely received the publicity or the glamour of the mountain mines.

Winnowing is the fundamental dry-wash method. It involves screening out all the coarse gravel, then placing the 'fines' in a blanket and tossing them into a good breeze, one man at each end of the blanket. The lighter particles are blown away by the wind, while the heavier particles fall back onto the blanket. The weave of the blanket tends to catch and hold onto fine gold.

A *dry washer* is a device that uses a trough similar to a sluice box in construction. The riffles area of the dry washer is placed in an upright position at a sharp angle perpendicular to the trough. The dry gravel and sand is on a slight downward slope. A shaking, vibrating action by means of a mechanical, manual or motor-driven crank causes the material to feed over the riffles, settling the heavier gold to the bottom of the trough. It is essential in both winnowing and dry washing that the material be thoroughly dry and disintegrated, having first separated or pulverized the coarser material. Damp gravel obtained below the surface material should be dry throughout to maximize the recovery process. There are a number of dry washer designs available from mining supply houses or one can be built in the home workshop.

The traditional method of panning can be

A "crevice tool" or long screwdriver is used to thoroughly clean out cracks and fissures. Pan the accumulated material.

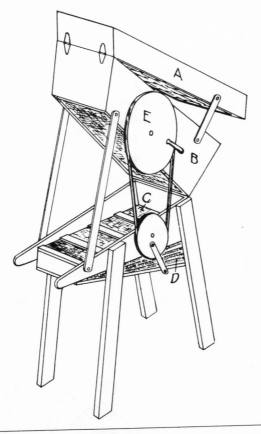

Dry Washer.
(Courtesy, California Division of Mines and Geology)

used by taking along a large old-fashioned wash bucket and a couple of five gallon cans of water and working the pan in a large bucket of water.

It should be noted for those not familiar with desert areas that temperatures, depending on altitude, can be severe with little natural cover to shield or hold the heat. Daytime summer heat can be deadly, as well as the exposure to cold winter nights. Perhaps even more than in other places, the necessity of leaving word with a friend on where you plan to go and when you expect to return is important since it takes considerably less time for a situation to turn critical. Proper clothing, including a hat to shield you from the sun's heat and glare, a small survival pouch (see details, page 95), and an adequate supply of drinking water are all essential.

Operation of a Sluice

Today many prospectors make use of an abbreviated version of the '49ers 'sluice-box'. Its use is called *sluicing*. It is a relatively short trough with a flat, smooth forward end and a riffle section. The sluice generally is placed right in the stream itself and works on the same principle as the flow of water along the bottom of the stream bed. A sluice can easily be made with basic tools from your workshop and several sections can be made for easy transportation. Several sizes are available in

Two miners operating a modern sluice box. Note the sieve used to sort the larger rocks.

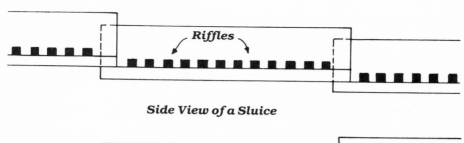

Side View of a Sluice

Top View of a Sluice

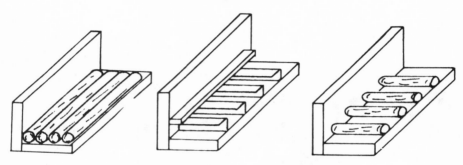

Various types of Riffles (one side cut away)

lightweight aluminum from various prospector supply houses. Generally a man can successfully pan anywhere from one-half to one yard of gravel per day. However, using a sluice one person can process six to eight times more material than by panning. It's a miniature 'stream' for purposes of separating the gold from the other river debris.

Sluices are generally four to six feet in length, six to twelve inches wide, with sides six inches high. Like the Long Tom, it is a three-sided, open-ended trough, fitted with a series of perpendicular riffles, cleats or obstructions every few inches which are set on a piece of burlap or indoor-outdoor carpeting.

In order to operate a sluice, place it in a rela-

(Drawing courtesy of California Department of Mines and Geology)

tively fast-moving current close to the bank where you will be working. Tip its head slightly higher than the tail so that the excess material will exit and disperse into the stream. If there is insufficient water running through the sluice, a wingdam made of a dozen or so good-sized rocks (see illustration) can be arranged at its head to divert additional water into the sluice. Generally a long, heavy rock is placed on top so that its weight will hold the sluice securely in place. Rocks can also be used under the head or tail of the box to regulate its pitch and the volume of water necessary to separate and classify the dirt fed into the hopper-end.

Set the sluice-box into the stream so that the water is running through it at a rate fast enough to carry most of the material the full length of the box. Take a pail full of material from a test area (upper end of a gravel bar, lee side of an obstacle, etc.) and begin to slowly pour handfuls of that material into the upper end of the sluice, breaking up the material with your hands to loosen and separate any clay, roots, etc., and sorting out any

Hopper end of sluice box is fed auriferous material for classification. (Inset: Eddy currents are formed behind riffles to trap gold.)

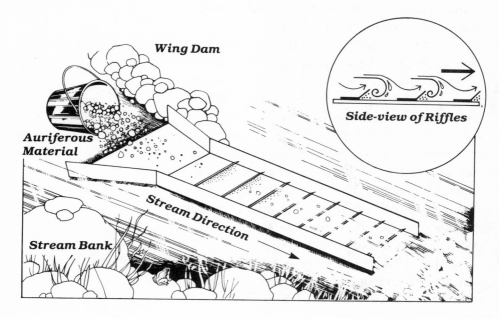

large rocks. Regulate the water passing through the sluice so that enough water keeps the riffles from becoming clogged.

After a while, a build-up of the heavier material and black sand will begin accumulating on the downstream side of the riffles. Regulate the angle of the sluice to pass all or most of the coarser material, letting the black sand (iron pyrites) gather behind the riffles. A few BBs can be dropped in the head to test the water action and obtain the optimum angle. These should ideally end up in the first two or three riffles, as will most of your gold. If they pass much beyond the third or fourth riffle either the water flow through the sluice is too swift or the angle too sharp, and much of your fine gold will be lost. Be sure you break up the clay, if any, since it has a strong tendency to hold on to any fine gold.

The black sand and any gold that will start to build up behind the riffles is doing so because of eddy (contrary) current that is formed behind the man-made riffle or cleat (obstacle). Continue to slowly put in and thoroughly mix the material at the head of the sluice until the riffles are full of 'concentrate'. When this happens (anywhere from two to three buckets of material to a dozen depending on the amount of iron pyrites in the soil), you're ready to 'clean up' and pan down the concentrate to remove the gold.

The purpose of the sluice box is to allow you to remove a much greater amount of *'overburden,'* with the aid of the stream, than you would by panning. It's generally wise to sort out larger rocks by hand or through a coarse screen prior to putting the material into the head of the sluice. This practice helps you avoid putting your hands into the area of the sluice to clear these, an action which can sometimes momentarily disrupt the eddy current and wash away the finer gold.

You can also improve the operation by fitting the sluice with a primary section constructed to fit in the 'mouth' of the sluice. This is called a "puddling box" and can be fabricated and fitted to the forward section of either a sluice or Long Tom. It's useful when muddy or clayey material is encountered. It allows this material, which is a notorious 'gold-robber,' to be thoroughly saturated, loosened, and broken prior to passing into the main sluice section.

Clean-up

In order to 'clean up,' stop the flow of water entering the sluice, and carefully remove it to the nearby bank. Remove the riffles, screen, and carpet and wash the concentrate with a bucket of water poured slowly into either a second bucket or your gold pan. *Thoroughly* wash the burlap or carpeting into the bucket until all gold particles are freed.

Your concentrate will be a considerably larger quantity than you would normally find after working down a single pan by the traditional method. The reason is that you have processed several times the amount of material with the sluice box and consequently have a proportionally larger amount of concentrate.

Pan down this concentrate as you normally would, taking each phase slowly. You're working with the heaviest material normally found after the first few steps in an initial panning operation.

When constructing a sluice box, or in an effort to 'improve' the commercially available type, some people replace the carpet material and fit the bottom with a piece of artificial grass or turf which is thought to trap more fine gold. After a season of use, the lining in the sluice bottom (carpet, burlap, turf) can be burned in a large steel gold pan and panned to recover the accumulated fine gold dust.

> **SLUICING.**
> To the right a company of miners are "sluicing;" those at the upper end are throwing in the pay dirt, and the man at the lower end is tending the sluice. Several lengths of sluice-boxes, or troughs with the ends out, supported by tressels, form the sluice; across the bottom, inside, are riffles or false bottoms, to save the gold; a stream of water being turned down, the gold is separated from the dirt, which is washed out

Cradle rocking on the Stanislaus. "A Book For Travellers and Settlers" by Charles Nordhoff, 1873.
(Courtesy of The California Historical Society, San Francisco)

Rocker Box

An alternative method to sluicing is by the fabrication of a *rocker-box* or *cradle*. The general dimensions of a rocker-box are similar, although each miner in the old days had his own variation. This method was used in the early mining operations in Georgia and the Carolinas. The function of this method is similar to that of the sluice with its own distinct advantage: it uses much less water than either a Long Tom or sluice.

Generally, for ease of operation, two or three people worked in a team. The device resembles an infant's cradle, as its name implies. It consists of a sluice-type open trough with riffles or cleats fashioned along the bottom which has a gradual sloping downward angle. It sits on a set of "rockers" fore and aft and at right angles to its length. At the forward end a box is attached with a coarse screen or sieve to separate the larger rocks, below it an "apron" of carpet is placed at a 30–40 degree angle.

The raw material is fed in the top over the screen which separates the various materials. Water via a bucket, diverted stream flow, or pump, pours into the top thereby pushing the material through the screen where it hits the apron secured beneath and then down to an arrangement similar to a short sluice box. The flow of water, mud, smaller rock and gravel-like material will then pass over the riffles. One man may be digging, a second rapidly pours water over the dirt in the box section and one of the two, or a third partner, 'rocks' the cradle with a jerky motion.

The separation or classification will settle out the gold as in the sluice operation. Again, as the riffles become full, the concentrate is taken out and panned to recover the gold.

The rocker which hasn't been widely used in recent years is experiencing a comeback. Its traditional styling, functional design, and its particular advantages are becoming more widely known. The average miner can pan up to a half cubic yard per day while a rocker operator can process four to five yards per day with less strenuous work. A rocker-box is portable, relatively lightweight and a very efficient hand-operated machine. Since it uses very little water it can be operated in semi-arid areas and even in desert areas with a recirculating water supply.

Long Tom

If conditions warranted its use in the old days a *Long Tom* was employed. It required a steady supply of water and usually two or three men to work it. A Long Tom was a three-sided inclined

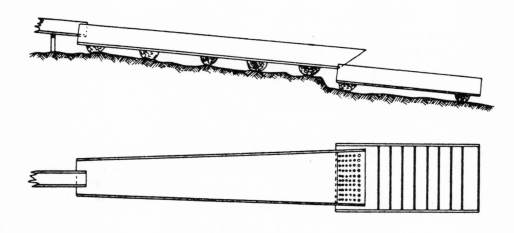

Diagram of a Long Tom.
(Courtesy, California Division of Mines and Geology)

trough that consisted of two sections. The upper end was approximately 10–12 feet long, built wider at the top than the bottom. The lower half of this first section was made of perforated sheet iron that acted as a sieve. The second section, called a riffle-box section, was usually half the length of the sluice and contained cleats or riffles across its bottom. In operation, material was shoveled into the top section of the trough and flushed down its length by a continuous flow of water directed into it. The sieve strained the coarser material while permitting the finer material and gold to pass through into the riffle-box where it became trapped.

Even this method had its limitation in terms of efficiency and an ingenious 19th century miner decided to modify it further. So was born the *Under Current Sluice*. A wide flat second sluice was placed under the regular sluice. It received its feed from a grating or screen placed in the bottom of the upper sluice and drew off the finer gold so it could be treated in a quieter current, with the larger coarse material and more turbulent water passing overhead.

Hydraulic Concentrator

One of the most novel devices I've used is a combination of the best features of a rocker, Long Tom, and sluice-box. It also doubles as a dredge! The unit is constructed of lightweight aluminum and is powered by a small, powerful two-cycle engine combined with a centrifugal pump. It can be used either to process material located at a distance from flowing water or, in its alternate arrangement, the nozzle/hose assembly is used to suck up gravel from the stream bed.

In an operation conducted away from an immediate supply of water, the engine/pump is placed at a convenient water source to draw water into the pump and deliver it through a long dis-

Hydraulic Concentrator. As illustrated, unit is set up to operate as a suction dredge.
(Courtesy, Keene Engineering Co.)

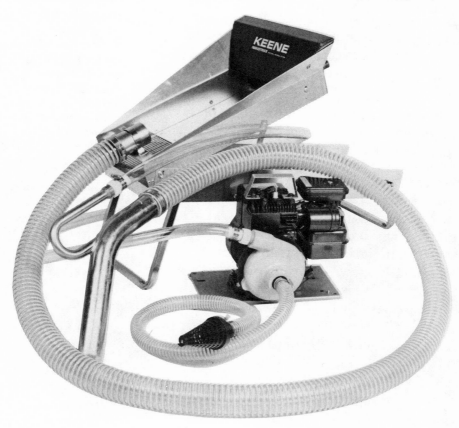

charge hose to the classifier/concentrator unit which is located at the area you are working. Gravel is shoveled into the upper end (hopper) of the concentrator unit, and water from the discharge hose washes the material. It then passes through the grizzly bars to separate the larger rocks and is classified in the lower (riffle) section.

As a dredge, the unit is set up at the stream bank. A suction hose and pressure hose combination are substituted for the discharge hose. The water passing from the water pump through the pressure at the nozzle causes a vacuum in the suction hose which is then used to suck up the potential ore-bearing gravel. This material again flows into the hopper, through the grizzly bars and then is 'sorted' in the riffle section. The gold is trapped in the same manner as in the other devices described earlier. Breakdown of the riffle section, underlying carpet, and cleanup is basically the same as for a rocker or sluice-box.

This unit which is capable of processing a considerable amount of material offers the operator the flexibility of owning one piece of equipment that is suited for both bench and placer operations.

Small backpack dredges offer portability for getting into those rich virgin areas. *(Courtesy of Treasure Emporium)*

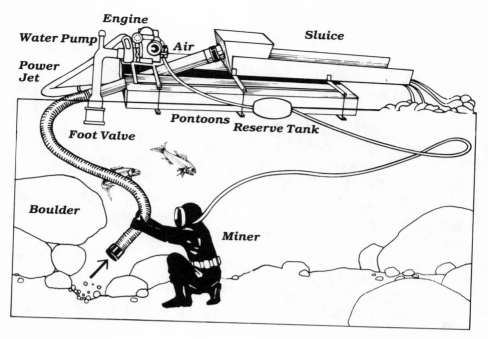

Typical setup of a recreational dredge showing miner directing suction hose.

Dredges

The most recent 20th century recreational method for the retrieval of stream placer gold is *dredging*. This method consists of a mechanical pump operated by a gas-powered engine that creates a vacuum. The whole apparatus is floated upon either a pair of large truck inner tubes or a set of pontoons. The vacuum hose is directed to suck up material from the stream bottom and the mixture of water, rock, sand, gravel and gold runs through a sluice where the gold is separated by the riffles.

These dredges generally operate with a several horsepower motor and come in a wide variety of styles and sizes moving quantities, at optimum, which range from one to thirty yards of gravel per hour, depending upon the horse power, hose diameter, altitude, and design of the dredge. The advantages are obvious, and although it would seem at first to be considerably less work than the pick, shovel, and pan method, there is a substantially larger investment in equipment, and

Author descending with dredge suction hose.

many obstructions that are larger than the capacity of the suction nozzle opening must still be moved manually.

There are several brands of 'portable' or 'backpack' dredges on the market that can be useful in sampling areas prior to committing to heavier equipment. They generally have an intake hose size of 1½ to 2½ inches, and the individual components can be broken down into a package carried by one man. Sufficient capacity for a *full-time* operation generally can't be generated until you trade up to a 3–5-inch model, thereby operating at a rated volume of 8–16 yards per hour of 'classified' material under ideal conditions.

The medium size dredges (3–5 inches) are the most popular because they are relatively lightweight and of moderate cost. Plan to buy or order one in the 'off season' as suppliers have recently not been able to keep up with demand as the price of gold climbs and popularity of the hobby increases.

Depending on the dredge's size, the operator's preference, and other variables (local restrictions, time of year, weather conditions), the operator can stand in the water or use a mask and snorkel while directing the suction hose. Deeper water operations require the use of air and compressor units are employed as a supplement to the power equipment. Air lines are used in lieu of scuba tanks for ease of movement, duration, and logistics of resupply. The latter method involves an additional investment in equipment for air compressors, reserve tank, hoses, harness, regulators, wet suits, etc., to operate under the surface of the water for several hours at a time. Those afflicted with the more advanced symptoms of 'gold fever' will eventually, and possibly quite rapidly, trade up to this method of pursuing their hobby since the rewards are proportional to the effort.

In principle, the basic elements for selecting a site are the same; that is, deciding on the optimum location where gold would "logically" have settled and drawing as much material, as is feasible in a given period of time, up the suction

Nose of suction hose sucks up gold-bearing gravels at stream bottom.

Large dredges like the one pictured were used to mine vast quantities of low-grade river placer deposits. *(Courtesy of Old Mint Museum, San Francisco)*

tube, through the baffle screen and over the riffles to classify and separate the gold from the other river material. There are several detailed books on dredging for gold should this method be of interest.

Metal Detectors

Electronic prospecting refers to locating nuggets and concentrations of black sand with its related deposits of gold by using a metal detector specifically designed for this purpose. It can be an end in itself or used as an adjunct to aid in selecting a site for any of the above methods. Additionally, it can be used in "hard rock" mining to locate the source of particularly rich veins.

Either the Beat Frequency Oscillator (BFO) or the new Very Low Frequency (VLF) type detector is used. These are employed because of their ability to produce true readings. The VLF has the ability to cancel out ground mineralization while still penetrating to satisfactory depths. Small (5-inch or less) detector search heads specifically designed for nugget hunting are offered by many manufacturers. This is a must to locate the smaller nuggets and offers maximum movement in the tight spots most likely to trap gold.

If you are interested in this method of prospecting, check the several brands that specialize in "nugget shooting." Generally, units are available in hip-mount and lightweight backpack models which add to their flexibility in the field.

Metal detectors can be an invaluable aid in desert dry washes. They eliminate the need to pan countless acres of dry, unproductive ground. By scanning placer areas in dried-up stream beds and gullies you can locate prime areas for using other equipment and locate nuggets buried under layers of sand and gravel. It is the fastest, most practical way to cover the most ground.

An advantage of the plastic pan during electronic prospecting is its use to isolate material when a reading from your detector looks promising. After a response from the metal detector's search head, a shovel is pushed as deeply as possible in the gravel to get under the source that activated the detector. Place your shovelful of paydirt into the pan and check the pan with the detector, repeating with more dirt and gravel until you discover your target.

Using a metal detector to locate deposits of black sands (magnetite) which generally also contain gold deposits. The search head can be used either on land or in shallow water.

CHAPTER V

Crescent of Gold

At some point you will undoubtedly ask yourself, "How do I know if this is real or fool's gold?" In spite of most popular misconceptions, there is a distinct difference between the two.

Fool's gold, usually yellowish iron or copper pyrite or yellow mica, is by comparison considerably lighter in weight and will float and/or swirl with the water in your gold pan. Gold, because of its specific gravity (19.2), will move very reluctantly. Fool's gold glitters and sparkles when held in the sun; gold 'glows' and its deep yellow coloring remains so even out of direct sunlight. When touched with the tip of a knife, fool's gold, which is brittle, will flake or break. Pyrites when struck with a hammer will shatter. Gold, however, is relatively soft and malleable and can be indented, scratched and cut with a knife blade. Iron pyrite is considerably harder than gold and will not be scratched by the tip of a knife blade. It will not react when placed in nitric acid (careful, highly caustic!) while fool's gold will foam, smoke, and finally disintegrate. The best test, however, is to see some real gold. Once you have, you'll see the vast difference and never confuse the two again.

Amalgamation

During the course of your sluicing and panning experience you'll pan down to the point of picking out the larger mini-nuggets and flakes. Eventually you'll have only very fine flour gold remaining which is difficult to isolate and pick up. Depending on your level of experience and the richness of the area, you may be satisfied only with the pieces you can pick up with your tweezers, discarding the minute 'dust'. If, however, you would like to accumulate finer gold as well, there is an old-time technique that's fairly effective. It requires the use of mercury and is called *amalgamation*. You'll need a small vial of mercury, a thin chamois cloth, a medium-sized white potato, some wire and tin foil.

The affinity of gold to adhere to the mercury while rejecting most of the black sand makes this process possible. To conserve time it's best to take the material you intend to treat and accumulate it until the end of the day. The concentrate should be panned down as much as possible then put in another vessel, such as a small plastic bowl, to avoid leaving any residual mercury on your pan.

Add the mercury to the bowl of concentrate, about a teaspoon or so will do (it'll take some trial and error to judge the right amount). Gently swirl the bowl, letting the ball of mercury roll over the surface of the black sand. After you're satisfied that it has come in contact with *all* the loose gold, roll the impregnated blob onto the chamois skin which has first been soaked in water. The *amalgam*, or mixture of mercury and gold, can now be separated by closing the top of the chamois skin and turning and squeezing it to press the mercury through the pores of the skin. This should be done over another bowl to retrieve the mercury. The flour gold will not pass through the cloth. All that remains is a small 'bead' of gold.

It is essential at this point to emphasize that mercury is a deadly poison and must be treated

Osterberg's "Quick Gold" separator

with great care. The fumes are deadly and it must always be used *outdoors*. Contact with the skin should also be avoided and it *must not* be used at all if you have any open sores or cuts on your fingers. *Always use in a well-ventilated area.*

There is an alternate method which is popular and practical when larger amounts of black sand are accumulated such as in large sluicing or dredging operations. This method employs the use of one of the many small rotary tumblers on the market offered by lapidary manufacturers and retail stores (see list). They are extremely effective in recovering minute particles of 'flour' gold.

Generally the tumbler is used first to agitate the contents of black sand. A small amount of caustic soda is added to clean the impurities from the gold and permits the mercury (added in the second phase) to more easily adhere to *all* the gold particles. After several hours of tumbling, the mercury-impregnated gold is placed through the chamois as in the manual method. Specific directions accompany most manufacturers' equipment.

Next you'll have to remove the residual mercury from the gold. One method is to cut a potato in half lengthwise. Scoop a small pocket out of the white center large enough to hold the gold bead. Put the halves together, secure with wire and wrap in several folds of aluminum foil. The potato is then placed in a hot open *outdoor* fire to bake. *Stay away from the fire to avoid inhaling the vapors.* After an hour or so, withdraw the potato. The mercury will have been absorbed into the white meat leaving a gold lump in the pocket. *Burn, bury, or properly dispose of the potato where no person or animal will eat it.* Do not permit the bowls used in the amalgamation to come in contact with any food stuff. This old-time method is for historical information only and for safety reasons it is *not* recommended.

Gemstone's amalgamator operates on a sonic system, reclaiming fine gold trapped in heavy sands. *(Courtesy of Gemstone Equipment Manufacturing Co.)*

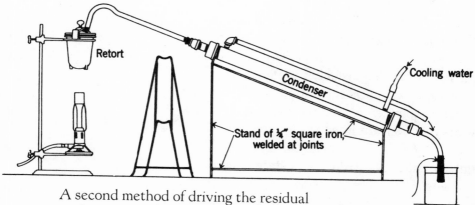

Retort.
(Courtesy, California Division of Mines and Geology)

A second method of driving the residual mercury from the gold saturated amalgam is a little more sophisticated and is called *retorting*. The *retort* works on the same principle as a still, that is, heat is used to drive off the mercury by means of distillation. The retort consists of four elements: a heat source such as a propane torch to heat the amalgam to 675 degrees, a tightly closed container to hold the amalgam, a long metal pipe which is enclosed in a water jacket to condense the vaporized mercury and a vessel at the terminal end to collect it.

Retorting is relatively simple, but the unit must be properly constructed and operated to assure that none of the gases escape. Again, it must be emphasized that the equipment must be used *outdoors* to avoid the possibility of inhaling any mercury vapors. Therefore, it's important to purchase rather than build a proper retort from a mining supply house and follow their instructions carefully for its safe operation.

Significant amounts of flour gold are contained in the black sand. However, unless you are accumulating large amounts of *concentrate,* such as in major sluicing or dredging operations, the use of mercury will not usually justify the cost. At all times, great care must be taken while handling the mercury.

If you are planning a larger operation, however, it *definitely* pays to hang onto your accumulated black sand. Serious prospectors are aware of its value and collect every ounce for further processing, usually through mechanical rather than chemical separation. There have been several equipment developments in the last few years that warrant interest.

One device separates the black sand/gold by putting the concentrate into a state of liquid suspension, thereby isolating the gold by specific gravity. The cylindrical apparatus allows a controlled velocity of water to flow into its base section, while the concentrate is introduced into the top. A "quicksand" of suspended material floats in the clear acrylic tube, the gold drops into a trap in the bottom of the unit, while the lighter waste colloids are siphoned off.

A larger scale method employs an adaptation of the Archimedean screw principle of raising water (and concentrate) by means of a spiral. By adjustment of the spiral wheel's angle and spray valve, all but the heaviest elements are flushed away, thereby recovering upwards of 90% of the fine gold through the wheel's central collecting point. These machines are set up and left to do their work while the miner goes about his other operations.

The Archimedes Spiral Separator.
(Courtesy of Gemstone Equipment Manufacturing Co.)

Staking Your Claims

If you have found a spot that *proves* to be good and you want to make an investment in time to expand the operation, you will want some assurance that someone else isn't going to start where you left off between trips. If the area is on public land, in most cases administered by the Bureau of Land Management (BLM), you can usually "stake a claim" to the mineral rights on that parcel. The governing laws vary from one state to another and within districts of the states. Dredging 'seasons' also vary to avoid any interference with fish migration and spawning seasons in certain rivers and streams. Specific information regarding details and forms for staking a claim can be obtained from the BLM office having jurisdiction over the area you are prospecting. I've listed those you are most likely to use at the end of this handbook.

Land status maps can be obtained for most areas from the local BLM office. These outline in broad forms the areas held in public domain within the area covered, and are keyed accordingly. There are, however, a number of private parcels which may not be shown because of the size of the scale. A specific area can usually be inquired about at the local ranger station. Forest rangers intimately know the terrain within their area.

Generally any U.S. citizen can "locate" and file a mining claim entitling the prospector to the mineral rights of a certain tract, usually 20 acres for an individual. The claim process usually has at least the following four elements:

1. A discovery of gold or other mineral within the boundary of the claim.
2. Erection of a monument and notice of location with description of its boundaries.

3. Clearly making the boundaries at the four corners.
4. Recordation of a true copy of the location in the appropriate county office and BLM.

The notice of location usually contains at least the date, county, township, state and name of the claimant. The filing itself contains the same information along with a physical description of the boundaries.

Once you have located, recorded and filed a claim with the necessary fees, marked the claim with its name, date of filing, and the name and addresses of the "holders," and meet all other local and federal requirements, you have established possessory rights to the minerals within its boundaries.

A non-patented claim entitles the holder(s) to the mineral rights but nothing else. The public can still hunt, fish, hike, and swim there as long as it is not directly interfering with the mining operation. Annual "assessment" work is required in the performance of a specified minimum amount of mining work to keep the claim active and valid, with the necessary documents being filed in the form of a statement of proof of labor.

As stated earlier, this is a generalized capsulation of the steps you will need to follow. Each area is governed by at least one, and usually several governmental agencies, either federal, state, local, or any combination. Begin with the local BLM and ask for their circular on current regulations. Be sure to ask what other agencies are involved (usually the Department of Fish and Game for dredging operations) and get the location of the closest ranger station within the area you wish to prospect.

There is another point that must be stated. As the price of gold escalates to and perhaps over the $1000 an ounce mark, there will be more and more territorial "rights" being established to sections of potentially profitable gold areas. In the mountains and deserts as elsewhere, there are some people who may be very aggressive about protecting what they think of as "theirs," either real or imagined. Most of these people have a legitimate right to the ground they're protecting and have spent a lot of time and money developing it. They've also probably had the "fever" long before gold reached its current price levels. So respect all posted areas. There is still an awful lot of open ground in our national forests and other federal, state and county land. Some of it may not be right beside a paved four-lane highway but then again it's the most easily accessible spots that have been worked the hardest by most people. Research and enjoy the secondary roads and trails. That's where the greatest rewards lie. It'll also enrich your overall sense of experience and appreciation of the area you're prospecting.

TUNNELING.

CHAPTER VI

Head for the Hills!

The recreational recovery of gold is a hobby enjoyed by thousands for the pure pleasure of being outdoors in country you might not normally visit. It's a trip into the past to an earlier time in this country's history, to the towns that were formed because of the richness of the surrounding soil. A day, a weekend, a vacation, or a camping holiday can be formed around this activity—the total experience being the sum of its parts. However, unlike other outdoor activities such as fishing and hunting, you or the people you mention your hobby to will invariably ask: What is it worth? How much did you find? Did it pay for the trip? I guess this is natural because of the universal awareness of the 'value' of gold.

You have to consider a couple of things in your own mind regarding its worth. Are you in it for the sport or for profit? If you are doing it for the fun of it, a guide to its estimated value would be helpful. On the other hand, if you're pursuing this as labor for its monetary remuneration because of the potential profit, you'll soon escalate your involvement into a considerable amount of equipment and capital, including precise instruments to measure the amount of gold recovered.

For the recreational prospector, some examples may be helpful. A day in a good area, using a sluice, can produce a yield from several grains to a couple of pennyweight, one pennyweight being 1/20 of a troy ounce. At today's prices that can add up pretty quickly. A Kruggerand, which is one troy ounce of gold, is approximately the size of a half dollar but slightly thicker. If you mentally cut up the surface area of that coin in fractions of a quarter or less, you can 'ballpark' the amount you've found. Inexpensive scales ($20-25) that are fairly precise for the price are available at supply houses.

Once you begin finding pieces in the nugget category, technically any piece of gold that weighs more than a grain (480 to the troy ounce) but generally anything from a half pennyweight on up, you will begin to appreciate more of a return than from "fine" gold. These pieces can be used for jewelry in the form of earrings, pendants and ring mounts, and command prices from a minimum of twice the spot price on up several times, depending on the size, texture, and character of the nugget. Pieces that have quartz mixed in with the

Off to the "diggings." *(Photo courtesy of Southern Pacific Transportation Co.)*

metal are particularly desirable, with varying value in proportion to the formation, tint, and amount of quartz. These specimen pieces are highly prized by collectors and are somewhat rare.

The grains and dust you find can be displayed in the small vials available at variety stores, encapsulated in small, clear plastic spheres available at jewelry stores, and hung as pendants. The "fines" can also be melted together to make a unique freeform piece of jewelry.

If you accumulate enough gold and wish to sell it, contact one of the refiners of precious metals in the Yellow Pages. Be sure to hold aside the "nugget" pieces. They will not be worth any more to the refiner whose only concern is to melt them together with the rest into large castings for resale.

Once you begin your quest, in all likelihood it will open and expand your interest into an exciting new endeavor. It will spur your interest in the history of our recent past, the geology of the area, and the details of the hills and mountains that surround us all. The search for that shining, elusive metal will be both exciting and frustrating but through it all will be a rich experience you will not soon forget, nor will it leave you unchanged. Keep in mind the pursuit is almost as much fun as the reward. Search for it, work for it, but above all, enjoy it!

Good Luck!

A primitive outfit. (After A Sketch From Life In 1850, by J. W. Audubon, in the possession of his daughter, Miss M. R. Audubon.)
(Courtesy of The California Historical Society, San Francisco)

Chapter VII

Nickel Knowledge

In this closing section I've included a potpourri of practical information that should be helpful in your further understanding of gold and the steps leading to its recovery. Specific information and equipment should be obtained according to your individual requirements.

Survival Pack

The following is a suggested list of items that can easily fit in a small "surplus" canvas magazine or cartridge pouch, "fanny" pack or the like, in the event you get stranded. All will fit in a package no larger than 5x5x3 inches.

Compass	Energy Chocolate Bar
Snake Bite Kit	Pencil and Note Paper
Band-Aids	Fishing Hooks and
Safety Pins	50 Feet of Line
Salt Pills	Whistle
Water Purification	Signal Mirror
Tablets	Plastic Tube Tent
Matches/	15 feet Nylon Cord
"Metal" Match	Insect Repellent
Candle	"Space Blanket"

Don't trespass on posted or fenced property.
" prospect in national parks/monuments.
" go off into the boonies unprepared or unfamiliar with the terrain without letting someone know where you are.
" camp or prospect in dry washes if distant storms are prevalent; flash floods can originate from storms miles distant.
" enter any old tunnels or shafts.
" get discouraged!

Do check with the department of Fish and Game regarding local ordinances.
" have a game plan, survey the area prior to your trip to conserve time later on during your vacation.
" look for all key spots and clues as discussed.
" take spares of easily lost items (tweezers, vials, etc.)
" respect posted mining claims.
" look out for snakes sunning themselves.
" take plenty of water if not available at your favorite mining site and drink often to avoid heat exhaustion.
" count on your intuition. The subconscious mind can put together the ingredients and intangible evidence we've discussed for your "glory hole".

Specifications

Symbol......................... Au
Atomic Number..................... 79
Atomic Weight................. 196.967
Melting Point................. 1063°C
Specific Gravity.................. 19.2
MOH's Scale of Hardness............ 2.5

The weight of gold is always determined in Troy ounces as opposed to the avoirdupois scale used in most other daily commodities. The common denominator for the two scales is grains. There are 437.5 grains per avoirdupois ounce as opposed to 480 grains in one troy ounce.

Troy Weight

1 grain = 0.0648 grams
24 grains = 1 penny weight (DWT) = 1.5552 grams
20 pennyweight = 1 ounce (480 grains) = 31.10 grams
12 ounces = 1 troy lb. = 5760 grains = 373.24 grams
Laboratory Pure Gold is 1.000 Fine
Commercially Fine Gold is .999 Fine
24 Karat = 100% Pure
18 Karat = 75% Pure
14 Karat = 58% Pure
10 Karat = 42% Pure
U.S. Gold Coins are .9166 Fine (22 Karat)

	Wt/Grms	Grains Pure
Double Eagle $20	33.4370	495.0
Eagle $10	16.7185	247.50
Half Eagle $5	8.3592	123.75
Quarter Eagle $2.50	4.1796	61.875
One Dollar $1	1.6718	24.750

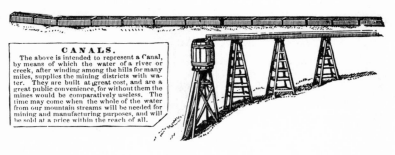

CANALS.
The above is intended to represent a Canal, by means of which the water of a river or creek, after winding among the hills for many miles, supplies the mining districts with water. They are built at great cost, and are a great public convenience, for without them the mines would be comparatively useless. The time may come when the whole of the water from our mountain streams will be needed for mining and manufacturing purposes, and will be sold at a price within the reach of all.

Mining Supply Resources

The following is a cross-section of some of the prospecting supply houses nationwide that carry mining and lapidary equipment. This is by no means a complete list nor does the omission of a dealer's name indicate that he is not reliable. Check your Yellow Pages under *Mining—Supplies* for a local dealer or write one of those listed below for their catalog.

Arizona

A & B Prospecting & Mining Equip.
   1735 W. Apache Trail, Suite 3
   Apache Junction, AZ 85220
Arizona Hiking Shack
   11645 N. Cave Creek Road
   Phoenix, AZ 85020
Gold Prospectors
   3342 W. Bell Road
   Phoenix, AZ 85023
Lucky Treasure World
   6005D W. Thomas Road
   Phoenix, AZ 85033
Prescott Gold Prospectors
   201 N. Cortez
   Prescott, AZ 86301
Treasure Shack
   2190 E. Apache Blvd.
   Tempe, AZ 85281

California

LoSierra Mining Supplies
   179 Palm Avenue
   Auburn, CA 95603
Auburn Detectors
   207 Werrow Street
   Auburn, CA 95603
PIONEER MINING SUPPLIES
   943 Lincoln Way
   Auburn, CA 95603
Teplow Drug Inc.
   404 E. Main
   Barstow, CA 92311
G. C. DeFabrizio & Associates
   16238 Lakewood Blvd.
   Bellflower, CA 90706
Frazier Lapidary Supply
   1724 University Avenue
   Berkeley, CA 94703
M & M Rock Shop
   40571 Lakeview Drive
   Big Bear Lake, CA 92315
Aurora Prospector Supply
   6286 Beach Blvd.
   Buena Park, CA 90620
Indian Valley Outpost
   Route 49
   Camptonville, CA 95922
Honan & Honan Mining
   2838 Highway 32
   Chico, CA 95926
Nuggets
   4018 Main Street
   Colfax, CA 95713
Goldsmith/Columbia Rock Shop
   22499 Parrots Ferry Road
   Columbia, CA 95310
Gold Nugget Detector Palace
   132 Cabrillo Street
   Costa Mesa, CA 92627
Yoho Downieville Gold Sales
   P.O. Box 431
   Downieville, CA 95936
Pro Sport Center
   508 Myrtle Avenue
   Eureka, CA 95501
Treasure Finders
   10209 Fair Oaks Blvd.
   Fair Oaks, CA 95628
The Prospector
   505 E. Broadway
   Glendale, CA 91205

Prospector's Hobby Shop
208 W. A Street
Hayward, CA 94541

Gold Lust Ltd.
18111 Mojave St.
Hesperia, CA 92345

Fumble Fingers
1027 Brown Avenue
Lafayette, CA 94549

Wyatts Detector Sales
42436 22nd West
Lancaster, CA 93534

Prospector Claim
2124 1st Street
Livermore, CA 94550

Treasure Hunters Supply
514 N. Santa Cruz
Los Gatos, CA 95030

Washington Square Coin Exc.
523 Washington Square
Marysville, CA 95901

GEMAC
1797 Capri Avenue
Mentone, CA 92359

Ace Hardware
2197 Central Avenue
McKinleyville, CA 95521

Gold Nugget
1302 Ninth Street
Modesto, CA 95354

Aqua Tech Dive Center
1717 Yosemite Blvd.
Modesto, CA 95351

W. H. Haney Co. Gems Galore
240 Castro Street
Mountain View, CA 94040

Alpha Hardware
210 Broad Street
Nevada City, CA

Nevada City Mint
30 Main Street
Nevada City, CA 95959

Treasure Emporium
6507 Lankershim Blvd.
North Hollywood, CA 91606

Keene Engineering Inc.
9330 Corbin Avenue
Northridge, CA 91324

Marin Trophies
932 Grant Avenue
Novato, CA 94947

Oakhurst Books & Things
Oakhurst Shopping Center Road 426
Oakhurst, CA 93644

Allied Sales
966 N. Main Street
Orange, CA 92667

Gold Town Mining Supplies
3800 Feather River Blvd.
Oroville, CA 95965

Gold Inc.
6165 Skyway
Paradise, CA 95969

Cal Gold Ent.
2400 E. Foothill Blvd.
Pasadena, CA 91107

The Gold Shop
364 Main Street
Placerville, CA 95667

Hangtown Etcetera
433 Main Street
Placerville, CA 95667

Sports Unlimited
681 Main Street
Placerville, CA 95667

Olde West
1330 Market Street
Redding, CA 96001

Devore Mining Supplies
831-B Sweeney Avenue
Redwood City, CA 94063

The Prospector
3823 McDonald Avenue
Richmond, CA 94805

Riverside Stamp & Coin
6740 Magnolia
Riverside, CA 92506

Mother Lode Dive Shop
2020 "H" Street
Sacramento, CA 95814

Valley Trophies & Detectors
1021 S. Main Street
Salinas, CA 93901

Kopp's Detectors
5590 Elmwood Road
San Bernardino, CA 92404

House of Treasure Hunters
5714 El Cajon Blvd.
San Diego, CA 92115

San Diego Coin Excange
3812 Grim Avenue
San Diego, CA 92104

Binkley's Lapidary
  2202 Lincoln Avenue
  San Jose, CA 95125
Seabrook & Lapidary Supply
  135 Third Street
  San Rafael, CA 94901
Jimmy Sierra Treasures
  1566 Fourth Street
  San Rafael, CA 94901
Gemstone Equipment Manufacturing
  480 E. Easy Street
  Simi Valley, CA 93065
Value Giant - Payless Drug
  750 E. Mono Way
  Sonoma, CA 95370
Tahoe Treasure Cove
  P.O. 1120
  South Lake Tahoe, CA 95705
Wright's
  2405 N. Eldorado Street
  Stocton, CA 95204
Rockteria Lapidary Supply
  1664 Cravens Avenue
  Torrance, CA 90501
Aladdin Communications
  301 N. Azusa Avenue
  West Covina, CA 91793
A-1 Cycle and Clock Repair
  13772 Golden West
  Westminster, CA 92683
Brown's Metal Detectors
  2868 Carr Drive
  Yuba City, CA 95991

Colorado
H. Glenn Carson
  801 Juniper Avenue
  Boulder, CO 80302
C B Services
  5901 N. Federal Blvd.
  Denver, CO 80321
C & D Detection Ent. Inc.
  5885 W. 38th Avenue
  Denver, CO 80212
Rocky Mountain Metal Detectors
  734 South Xavier Street
  Denver, CO 80219
The Prospector's Cache
  25 West Girard
  Englewood, CO 80110

Florida
Ben's Coin and Gem Shop
  1023 Ridgewood Avenue
  Hollyhill, FL 32017
Crystal Gems
  8453 S.W. 132nd Street
  Miami, FL 33156
Darwin's Treasure Finders
  4478 Park Blvd.
  Pinellas Park, FL 33565

Georgia
Buck's Marine & Gold
  On The Square
  Dahlonega, GA 30533

Idaho
Q's Trophy Cabin
  3940 Overland Road
  Boise, ID 83705
Idaho Stamp & Coin
  3506 Rosehill
  Boise, ID 83705
The Rock Shop
  490 N. 2nd E.
  Mountain Home, ID 83647

Minnesota
Twin City Metal Detectors
  403 3rd Street
  Elk River, MN 55330
Metal Detectors of Minneapolis
  615 West Lake Street
  Minneapolis, MN 55408

Missouri
The Treasure Hut
  1315 N. Main
  Poplar Bluff, MO 63901
Plateau Detector Center
  9837 Kimker
  St. Louis, MO 63127

Nevada
Gold Prospectors Supply
  1441 Rand Avenue
  Carson City, NV 89701

New Mexico
Alaska Gold Mining
  925 Eubank N.E.
  Albuquerque, NM 87112
The Maisel Co.
  1500 Lomas Avenue N.W.
  Albuquerque, NM 87104

Roswell Treasure Center
#12 Monterey Shopping Center
1400 West Second Street
Roswell, NM 88201

North Carolina
Treasure House
3201 Cullman Avenue
Charlotte, NC 28213

Oregon
Traditional Gold Dredges
1613 N.W. Division
Bend, OR 97701
Prospectors Supply Co.
210 E. Pearl
Coburg, OR 97401
D & K Detector Sales
540 S.E. 10th
Hillsboro, OR 97123
Medford Coin & Prospecting Supply
408 E. Main
Medford, OR 97501
D & K Detector Sales
13809 S.E. Division
Portland, OR 97236
Le Bleu's Rock Shop
1810 N.E. Stephens
Roseburg, OR 97470
Applegate Miners
7388 Highway 238
Ruch, OR 97530

Texas
Sun City Coin & Stamp Co.
87 D Bassett Center
El Paso, TX 79925
Fireball Electronics
1541 18 Parkway
Odessa, TX 79762
Jarl's Treasure Supplies
1108 S. Shaver
Pasadena, TX 77506

Utah
T. R. Baker Sales
4190 W. 5500 South
Kearns, UT 84118

Washington
Alpha Faceting Supply Co.
1225 Hollis Street
Bremerton, WA 98310
Prof's Handicrafts
56 N. College Avenue
College Place, WA 99324
Doug's Mining Supplies
13501 100th Avenue N.E.
Juanita, WA 98033
Doug's Mining Supplies
14704 112th Avenue N.E.
Kirkland, WA 98033
Renton Coin Shop
225 Well Avenue S.
Renton, WA 98055
Placer Creek Prospecting
1620 W. Clark
Pasco, WA 99301
Prospector Ed's Gold Supplies
5263 Rainier Avenue South
Seattle, WA 98118
Lortone Inc.
Lapidary Equipment Manufacturers
2856 N.W. Market Street
Seattle, WA 98107
Pearl Electronics Inc.
1300 First Avenue
Seattle, WA 98101
Bowen's Hideout
South 1823 Mt. Vernon
Spokane, WA 99203
Wenatchee Prospecting Supply
35 S. Wenatchee Avenue
Wenatchee, WA 98801

Wisconsin
Don's Treasure Hunting Supply
N88 W16747 Appleton Way
Menomonee, WI 53051

Western Canada Distributor
Rocky Mountain Detectors, Ltd.
P.O. Box 5366, Postal Station "A"
Calgary, Alberta T2H 1X8, Canada

Notes

Bureau of Land Management (BLM) Offices

Alaska State Office
555 Cordova Street
Anchorage, AK 99501
(907) 277-1561

Arizona State Office
2400 Valley Bank Center
Phoenix, AZ 85073
(602) 261-3873

California State Office
Federal Building
Sacramento, CA 95825
(916) 484-4676

Colorado State Office
Colorado State Bank Building
Denver, CO 80202
(303) 837-4325

Eastern States Office
(All states east of Mississippi River)
7981 Eastern Avenue
Silver Springs, MD 20910
(301) 427-7500

Idaho State Office
Federal Building
Boise, ID 83724
(208) 384-1401

Montana State Office
(Montana, North and South Dakota)
Granite Tower Building
222 N. 32nd Street
Billings, MT 59101
(406) 657-6461

Nevada State Office
Federal Building
Reno, NV 59609
(702) 784-5451

New Mexico State Office
(New Mexico, Oklahoma, and Texas)
Federal Building
Santa Fe, NM 87501

Oregon State Office
(Oregon & Washington)
729 N.E. Oregon Street
Portland, OR 97208

Utah State Office
University Club Building
136 E. South Temple Street
Salt Lake City, UT 84111
(801) 524-5311

Wyoming State Office
(Wyoming, Nebraska, and Kansas)
Federal Building
Cheyenne, WY 82001
(307) 778-2326

Glossary

Amalgamation – The use of mercury to collect fine gold from the final washed down concentrate.

Atomic Number – A number representing the weight of one atom of an element as compared with an arbitrarily selected number representing the weight of another element taken as the standard.

Arrastre – A mechanical device consisting of a yoke and millstone usually turned by mule power in the early days of mining to crush rock and separate the gold.

Auriferous – Ground or material bearing or yielding gold.

Bar/Gravel Bar – A deposit of gravels and rock in a stream usually caused by the slower currents of water and associated with gold where present in rivers.

Bench – A terrace of gravel along the bank of a stream left by the action of the water in earlier times.

Claim Jumping – Stealing someone else's mining claim before it's been recorded or, nowadays, the use during the owner's absence.

Colors – The particles of gold amid the black sand left in the gold pan after washing.

Coyoteing/Coyote Holes – The practice of sinking a shaft or tunnel large enough for one man to crawl into while following the path of a rich "vein" of gold.

Dredging – The use of a subsurface hose powered by an engine to suck up auriferous material which is in turn separated in the sluice portion of the equipment.

Dry Washer – A mechanical device for separating the gravels and soil from the gold without the use of water.

Fool's Gold – Mineral sometimes mistaken for gold, made up of iron sulfide or copper-iron sulfides and often yellow or brass in color.

Glory Hole – A relatively small but concentrated pocket of gold.

Gold Fever – A disease, when contracted, having no known cure; periods of dormancy sometimes occur in winter months. Symptoms include a faraway look, a "need" to get into the mountains, an expanded love of the outdoors, and an endless search for the buried location of the pot at the end of the rainbow.

Grain – Originally derived from the weight of a single grain of wheat, the smallest unit in the system of weights and measures.

Grubstake – Supplying a prospector with "grub" and equipment for a percentage of his take.

Hydraulicking – The practice now outlawed or severly restricted in most areas of directing a high-pressure stream of water at benches and river banks to wash the material into sluices and "Long Toms" to recover gold.

Lode – A deposit or vein containing quantities of valuable mineral mined by digging a tunnel or sinking a shaft.

Long Tom – A large, usually wooden trough used in the early days of mining in which gold-bearing material and water were used to separate the gold from the gravels.

Motherlode – The main "lode" or vein of ore; the general region having a considerable quantity of auriferous material, referred to as the central mining district on the western flanks of the Sierra range.

Paydirt – Soil, gravel, ores, etc., rich enough in minerals to make mining profitable.

Placer – A deposit of gravel or sand containing heavy ore minerals including gold which has eroded from the surrounding mountains.

Quartz – A hexagonal crystalline material often colorless or milk-white and associated with gold "veins."

Rocker – Early invention in the general shape of a cradle which combined the aspects of a sluice with a section to sort out the rocks built in its forward section.

Stamp Mill – Any number of designs used to crush ore and separate the gold from the host material.

Tailings – The waste material from a former mining operation consisting of processed gravels and rock left as a waste product and often containing gold missed by the former operation.

Wingdam – Obstruction constructed to divert water to or away from a mining operation.

(Photo courtesy of Levi Strauss & Co.)

Mining Supplies

The following equipment are the basic "tools" of the modern prospector and can be ordered directly from : Sierra Trading Post, P.O. Box 2497, San Francisco, California 94126. Please order early as supplies become limited during the spring and summer seasons.

*Medium Size Gold Pan
Advanced design, high impact plastic with built-in riffles, textured surface, non-corrosive and lightweight. Approx. 14 inch dia. **Price: $5.95**

*Gold Pan Sieve
High impact plastic, 12½ inches in diameter, can be used with any type gold pan of approximately the same size or larger. Screens out waste gravel to cut panning time in half! **Price: $5.95**

Hydrosluice
Mini-sluice box with all the features of a professional sluice. Flared to gather water for efficient recovery of the finest gold. Equipped with fine gold trap that has a removeable screen and carpet – molded of "marlex polymer," one of the toughest known plastics. Approx. 15 x 36 inches – 4 lbs. **Price: $24.95**

Mini-Sluice
Light-weight aluminum construction, features removeable riffles, diamond expanded screen, and DuPont matting. Approx. 10 x 36 inches – 5 lbs.
Price: $34.95

Hand-Sluice
Aluminum construction with removeable riffles, diamond expanded screen for ultra fine gold recovery, lined with DuPont matting. Flared hopper for increasing velocity for heavier capacity. Approx. 10 x 52 inches – 10 LBS. **Price: $54.95**

*Gold Scale
Its unique design folds into a pocket container. This amazingly accurate scale will weigh from 1 grain to a maximum of 1 ounce. Includes 10 pieces of troy weights.
Price: $27.95

Effective ·1/1/84

Prospector's Special
Get started with the proper equipment! Kit includes medium-size gold pan, gold extractor magnet, magnifier, crevice tool, 1 oz. sample bottle, prospector's pick, classifier sieve - 10 lbs. **Price: $25.95**

Rocker-Box
Processes ten times the material you can by panning in the same amount of time and with a fraction of the effort. Excellent for recovery of fine gold in areas where there is a limited supply of water. Individually hand-crafted, features traditional styling, rugged construction, and portability. Shipping weight approx. 30 lbs. **Price: $155.00**

Hydraulic Concentrators
Model 163-3. Powered by our Model P-103 pump and engine assembly capable of lifting water as high as 200 feet above stream level and out several hundred feet. Equipped with 25 feet of pressure hose and P-7 foot valve and hose assembly. The concentrator is equipped with a gate valve for water control. Just dial the proper amount of water to best do the job. The concentrator folds up into a compact package for carrying and weighs only 17 lbs. Total unit weight is 49 lbs. **Price: $495.00**

Model 165-3. The ultimate in a gold recovery machine! We have adapted a 2½-inch dredge to our hydraulic concentrator. This is an ideal and most versatile machine, having all the qualities of a super fine gold recovery machine and the capabilities and production of a dredge. Equipped with the same basic equipment as the 163-3. Weighs 66 lbs.:*(as illustrated p. 77)* **Price: $595.00**

Backpack Dredge
Features 1½ hp engine and pump. Delivers up to 60 gpm and weighs only 9½ lbs. Equipped with 2 inch suction nozzle for extreme versatility in shallow water dredging. Extra wide sluice box for maximum fine gold recovery. Compact fold-up construction makes into a small package for back-packing. Capable of dredging up to 2 cubic yards per hour. The most economical, powerful and lightweight 2-inch dredge on the market. 38 lbs. *(as illustrated p. 78)* **Price: $439.50**

Many other mining supplies available including 3-4-5-6-8-inch dredges, etc. Please write us of your equipment needs:
Sierra Trading Post, P.O. Box 2497, San Francisco, CA 94126.

Effective 1/1/84

Gold Locations
of the United States

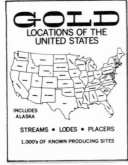

NEW!

Price $7.95
176 Pages - Soft Cover

Now, for the first time, all the known, possible gold recovery locations under one cover! More than 1,000 United States counties and gold producing districts, including Alaska, are discussed in this fabulous new research tool. Many of these individual locations have as many as twenty gold producing streams and lode sites listed.

Not only are the sites precisely located with landmarks and mileages, much other information is given about each district. Total gold production is given for each location, and in ounces wherever possible. This allows the prospector to determine whether enough gold was produced to make it worth his while to investigate the area.

But most important, Gold Locations of the United States, points out which streams and placers are replenishible. These facts are especially important in areas like the Mother Lode of Califorina where billions of ounces still remain in un-mined Tertiary Channels on which hydraulicking was closed down by the law in the 1880s. The book reveals which are still releasing gold to the streams by natural causes and tells where modern skin divers and dredge operators have the best luck.

Another useful bit of historical information that modern day prospectors will find very helpful is the listing of the type of gold that was produced in each location. When applicable the type and size of the dust dredged or sluiced is told. Those sites that produced nuggets of sizes from one-quarter ounce up are mentioned. Rare types of gold, like crystal forms are also stated.

Unlike most books of this nature this new volume also covers the East, where the first gold rush took place, and where gold is still being panned every day. All the know locations are listed, even those which have been closed down for a century.

This is assuredly the definitive work on gold locations in the United States and is now the most valuable research tool in thousands of prospector's libraries.

Include S.63 for postage

Prospector's Special

Get started with the proper equipment and save!
This kit has all the basic tools to get you started:

Only $25.95

Gold Pan
Classifier Sieve
Combination Pick/Mattock
Crevice Tool
Magnifying/Specimen Box
Black Sand/Gold Separator
1 oz. Gold Sample Vial

Please include $3.75 for postage & handling.

Order Form

Name of Article	Quantity	Price	Total
			$

Mail Order To:
Sierra Trading Post
P.O. Box 2497
San Francisco, CA 94126

Subtotal: $ _____
Calif. Residents add 6% Sales Tax: _____

Postage & Handling: _____

Total Enclosed: $ _____

Ordered By _____

Street Address _____

Town/City _____

State _____ **Zip** _____

Please print all information clearly so your order may be filled promptly.

Please send cashiers check or money order. Personal checks take approximately ten banking days to clear. No C.O.D.s

Shipping Instructions: All prices are F.O.B. San Francisco, California. Shipping cost is additional. For smaller equipment marked with an asterisk (*) send $1.50 for postage. For other items add $1.50 for the first pound and $.25 for each additional pound for postage and handling, or specify to send freight collect.

Due to changes of material cost beyond our control, prices are subject to change without notice.

Order Early! Allow approximately 3-4 weeks for delivery.

Many other mining supplies available including 3-4-5-6-8 inch dredges, etc. Please write us of your equipment needs.

TEAR HERE